Algebra II
Station Activities
for Common Core Standards

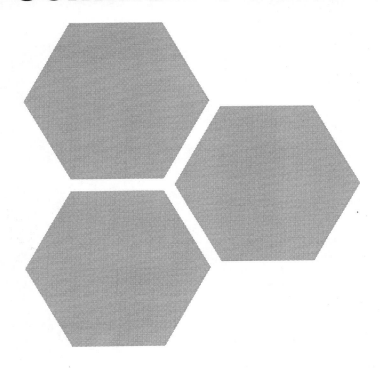

The classroom teacher may reproduce materials in this book for classroom use only.
The reproduction of any part for an entire school or school system is strictly prohibited.
No part of this publication may be transmitted, stored, or recorded in any form
without written permission from the publisher.

© Common Core State Standards for Mathematics. Copyright 2010. National Governor's
Association Center for Best Practices and Council of Chief State School Officers. All rights reserved.

1 2 3 4 5 6 7 8 9 10
ISBN 978-0-8251-6793-5
Copyright © 2011
J. Weston Walch, Publisher
Portland, ME 04103
www.walch.com
Printed in the United States of America

Table of Contents

Standards Correlations ....*v*
Introduction ....*vii*
Materials List ....*x*

Number and Quantity
    Set 1: Operations with Complex Numbers 1

Algebra
    Set 1: Factoring 10
    Set 2: Solving Quadratics 18
    Set 3: Linear Programming 27

Functions
Interpreting Functions 39
    Set 1: Quadratic Transformations in Vertex Form....................... 39
    Set 2: Graphing Quadratics 58
    Set 3: Piecewise Functions 69
    Set 4: Absolute Value Equations and Inequalities 82
    Set 5: Exponential Functions 91
    Set 6: Solving Exponential Equations and Inequalities 101
    Set 7: Polynomial Functions 118

Building Functions 132
    Set 1: Inverse Functions 132
    Set 2: Logarithmic Functions As Inverses of Exponential Functions 150

Linear, Quadratic, and Exponential Models 161
    Set 1: Arithmetic Sequences and Series 161
    Set 2: Geometric Sequences 170

Geometry
    Set 1: Conics 183

Statistics and Probability
    Set 1: Modeling 199
    Set 2: Sampling 210
    Set 3: *z*-scores 224

Standards Correlations

The standards correlations below support the implementation of the Common Core Standards. This book includes station activity sets for the Common Core domains of Number and Quantity, Algebra, Functions, Geometry, and Statistics and Probability. This table provides a listing of the available station activities organized by Common Core standard.

The left column lists the standard codes. The first letter of the code represents the Common Core domain. The domain letter is followed by a dash and the initials of the cluster name, which is then followed by the standard number. The middle column lists the title of the station activity set that corresponds to the standard, and the right column lists the page number where the station activity set can be found.

The table indicates the standards that are heavily addressed in the station sets. If there are other standards that are addressed within the set, they can be found on the first page of each set.

Standards	Set title	Page number
N-CN.1	Operations with Complex Numbers	1
N-CN.2	Operations with Complex Numbers	1
N-CN.3	Operations with Complex Numbers	1
N-CN.7.	Solving Quadratics	18
N-CN.9.	Polynomial Functions	118
A-SSE.2.	Polynomial Functions	118
A-SSE.3.	Factoring	10
A-SSE.3.	Solving Quadratics	18
A-APR.2.	Polynomial Functions	118
A-APR.3.	Polynomial Functions	118
A-REI.4.	Solving Quadratics	18
A-REI.4.	Graphing Quadratics	58
A-CED.2.	Linear Programming	27
A-CED.3.	Linear Programming	27
F-IF.2.	Piecewise Functions	69
F-IF.2.	Exponential Functions	91
F-IF.2.	Solving Exponential Equations and Inequalities	101

Standards Correlations

Standards	Set title	Page number
F-IF.2.	Polynomial Functions	118
F-IF.3.	Arithmetic Sequences and Series	161
F-IF.4.	Graphing Quadratics	58
F-IF.5.	Graphing Quadratics	58
F-IF.7.	Quadratic Transformations in Vertex Form	39
F-IF.7.	Solving Exponential Equations and Inequalities	101
F-IF.7.	Graphing Quadratics	58
F-IF.7.	Piecewise Functions	69
F-IF.7.	Absolute Value Equations and Inequalities	82
F-IF.7.	Exponential Functions	91
F-IF.7.	Polynomial Functions	118
F-IF.7.	Logarithmic Functions As Inverses of Exponential Functions	150
F-IF.8.	Quadratic Transformations in Vertex Form	39
F-BF.2.	Arithmetic Sequences and Series	161
F-BF.2.	Geometric Sequences	170
F-BF.3.	Quadratic Transformations in Vertex Form	39
F-BF.4.	Inverse Functions	132
F-BF.5.	Logarithmic Functions As Inverses of Exponential Functions	150
F-LE.2.	Arithmetic Sequences and Series	161
F-LE.2.	Geometric Sequences	170
G-GPE.1.	Conics	183
G-GPE.2.	Conics	183
G-GPE.3.	Conics	183
S-ID.2.	Sampling	210
S-ID.6.	Modeling	199
S-ID.9.	Modeling	199
S-ID.4.	z-scores	224

Introduction

This book includes a collection of station-based activities to provide students with opportunities to practice and apply the mathematical skills and concepts they are learning. It contains sets of activities for Number and Quantity; Algebra; Functions (Interpreting Functions; Building Functions; Linear, Quadratic, and Exponential Models); Geometry; and Statistics and Probability. You may use these activities in addition to direct instruction, or instead of direct instruction in areas where students understand the basic concepts but need practice. The Discussion Guide included with each set of activities provides an important opportunity to help students reflect on their experiences and synthesize their thinking. It also provides guidance for ongoing, informal assessment to inform instructional planning.

Implementation Guide

The following guidelines will help you prepare for and use the activity sets in this book.

Setting Up the Stations

Each activity set consists of four stations. Set up each station at a desk, or at several desks pushed together, with enough chairs for a small group of students. Place a card with the number of the station on the desk. Each station should also contain the materials specified in the teacher's notes, and a stack of student activity sheets (one copy per student). Place the required materials (as listed) at each station.

When a group of students arrives at a station, each student should take one of the activity sheets to record the group's work. Although students should work together to develop one set of answers for the entire group, each student should record the answers on his or her own activity sheet. This helps keep students engaged in the activity and gives each student a record of the activity for future reference.

Forming Groups of Students

All activity sets consist of four stations. You might divide the class into four groups by having students count off from 1 to 4. If you have a large class and want to have students working in small groups, you might set up two identical sets of stations, labeled A and B. In this way, the class can be divided into eight groups, with each group of students rotating through the "A" stations or "B" stations.

Introduction

Assigning Roles to Students

Students often work most productively in groups when each student has an assigned role. You may want to assign roles to students when they are assigned to groups and change the roles occasionally. Some possible roles are as follows:

- Reader—reads the steps of the activity aloud
- Facilitator—makes sure that each student in the group has a chance to speak and pose questions; also makes sure that each student agrees on each answer before it is written down
- Materials Manager—handles the materials at the station and makes sure the materials are put back in place at the end of the activity
- Timekeeper—tracks the group's progress to ensure that the activity is completed in the allotted time
- Spokesperson—speaks for the group during the debriefing session after the activities

Timing the Activities

The activities in this book are designed to take approximately 15 minutes per station. Therefore, you might plan on having groups change stations every 15 minutes, with a two-minute interval for moving from one station to the next. It is helpful to give students a "5-minute warning" before it is time to change stations.

Since the activity sets consist of four stations, the above time frame means that it will take about an hour and 10 minutes for groups to work through all stations. If this is followed by a 20-minute class discussion as described below, an entire activity set can be completed in about 90 minutes.

Guidelines for Students

Before starting the first activity set, you may want to review the following "ground rules" with students. You might also post the rules in the classroom.

- All students in a group should agree on each answer before it is written down. If there is a disagreement within the group, discuss it with one another.
- You can ask your teacher a question only if everyone in the group has the same question.
- If you finish early, work together to write problems of your own that are similar to the ones on the student activity sheet.
- Leave the station exactly as you found it. All materials should be in the same place and in the same condition as when you arrived.

Introduction

Debriefing the Activities

After each group has rotated through every station, bring students together for a brief class discussion. At this time, you might have the groups' spokespersons pose any questions they had about the activities. Before responding, ask if students in other groups encountered the same difficulty or if they have a response to the question. The class discussion is also a good time to reinforce the essential ideas of the activities. The questions that are provided in the teacher's notes for each activity set can serve as a guide to initiating this type of discussion.

You may want to collect the student activity sheets before beginning the class discussion. However, it can be beneficial to collect the sheets afterward so that students can refer to them during the discussion. This also gives students a chance to revisit and refine their work based on the debriefing session.

Guide to Common Core Standards Annotation

As you use this book, you will come across annotation symbols included with the Common Core standards for several station activities. The following descriptions of these annotation symbols are verbatim from the Common Core State Standards Initiative Web site, at www.corestandards.org.

Symbol: ★

Denotes: Modeling Standards

Modeling is best interpreted not as a collection of isolated topics but rather in relation to other standards. Making mathematical models is a Standard for Mathematical Practice, and specific modeling standards appear throughout the high school standards indicated by a star symbol (★).

From http://www.corestandards.org/the-standards/mathematics/high-school-modeling/introduction/

Symbol: (+)

Denotes: College and Career Readiness Standards

The evidence concerning college and career readiness shows clearly that the knowledge, skills, and practices important for readiness include a great deal of mathematics prior to the boundary defined by (+) symbols in these standards.

From http://www.corestandards.org/the-standards/mathematics/note-on-courses-and-transitions/courses-and-transitions/

Introduction

Materials List

- 12 index cards
- six-sided number cube
- algebra tiles
- calculators
- colored pens or pencils
- graph paper
- graphing calculators
- *optional*: computer station with graphing software
- rulers
- z-scores tables

Number and Quantity

Set 1: Operations with Complex Numbers

Instruction

Goal: To give students practice in adding, subtracting, multiplying, and dividing complex numbers; to help students recognize the relationship between complex numbers in radical form and numbers in $a + bi$ form

Common Core Standards

Number and Quantity: The Complex Number System

Perform arithmetic operations with complex numbers.

- **N-CN.1.** Know there is a complex number i such that $i^2 = -1$, and every complex number has the form $a + bi$ with a and b real.
- **N-CN.2.** Use the relation $i^2 = -1$ and the commutative, associative, and distributive properties to add, subtract, and multiply complex numbers.
- **N-CN.3.** (+) Find the conjugate of a complex number; use conjugates to find moduli and quotients of complex numbers.

Student Activities Overview and Answer Key

Station 1

Students race their partner to complete addition and subtraction problems involving complex numbers. Students check each other's work.

Answers

1. $3 + 8i$
2. $13 - 4i$
3. $22 + 5i$
4. $26 + 3i$
5. 5
6. $3a + 4gi$
7. $9i - 26$
8. 4
9. $7 - 7i$
10. $-4 + 17i$

Number and Quantity
Set 1: Operations with Complex Numbers

Instruction

Station 2

Students work with a partner to multiply complex numbers.

Answers

1. $-13 + 11i$
2. $-2 + 34i$
3. $1 + 8i$
4. $\dfrac{1}{2} + \dfrac{81}{4}i$
5. $-8 - 12i$
6. $13 - i$
7. $47 - 28i$
8. $50 + 10i$
9. 85

Station 3

Students work with groups to identify the conjugate c of complex numbers and solve division problems.

Answers

1. $c = 5 + 6i$

   $\dfrac{3}{61} + \dfrac{28}{61}i$

2. $c = 2 - i$

   $\dfrac{11}{5} + \dfrac{2}{5}i$

3. $c = 3 + 2i$

   $\dfrac{11}{13} + \dfrac{16}{13}i$

4. $c = 3 + 2i$

   $\dfrac{5}{13} + \dfrac{12i}{13}$

Number and Quantity
Set 1: Operations with Complex Numbers

Instruction

5. $c = 7 - 3i$
   1

6. $c = 2 - i$
   $\dfrac{13}{5} - \dfrac{14i}{5}$

7. $c = 5 - 3i$
   $\dfrac{12}{17} - \dfrac{14i}{17}$

8. $c = 4 - 2i$
   $\dfrac{1}{2} - \dfrac{i}{2}$

Station 4

Students work in groups to solve equations involving complex numbers, sometimes in radical form, sometimes in $a + bi$ form.

Answers

1. $8 + \dfrac{i}{2}$

2. $3 + 4i$

3. $-2 + 3i$

4. $-20 - 20i$

5. $\dfrac{2}{3} - \dfrac{2i}{3}$

6. $\dfrac{-8}{13} + \dfrac{12}{13}i$

7. 13

8. $\dfrac{-254}{221} + \dfrac{667}{221}i$

Number and Quantity
Set 1: Operations with Complex Numbers

Materials List/Setup

Station 1   none
Station 2   calculator
Station 3   none
Station 4   none

Number and Quantity
Set 1: Operations with Complex Numbers

Instruction

Discussion Guide

To support students in reflecting on the activities and to gather some formative information about student learning, use the following prompts to facilitate a class discussion to "debrief" the station activities.

Prompts/Questions

1. What is the square root of –1?
2. What is a complex number?
3. How do you find the conjugate of a complex number?
4. What is the product of a complex number $a + bi$ and its conjugate?

Think, Pair, Share

Have students jot down their own responses to questions, then discuss with a partner (who was not in their station group), and then discuss as a whole class.

Suggested Appropriate Responses

1. i
2. A complex number is one that can be expressed as $a + bi$, where a and b are real numbers.
3. If a complex number is expressed as $a + bi$, its conjugate is $a - bi$.
4. $a^2 + b^2$

Possible Misunderstandings/Mistakes

- Incorrectly multiplying polynomials
- Incorrectly finding the conjugate of a complex number
- Making simple arithmetical errors in adding, subtracting, multiplying, and dividing
- Not understanding the relationship between complex numbers in radical form and in $a + bi$ form
- Not recognizing that $i^4 = 1$ and $i^2 = -1$

Number and Quantity
Set 1: Operations with Complex Numbers

Station 1

Race your partner to complete the addition and subtraction problems. Show all work. When you have both finished, check each other's work.

1. $(1 + 3i) + (2 + 5i)$

2. $(3 + 7i) + (10 - 11i)$

3. $(18 + 3i) + (4 + 2i)$

4. $(16 + 2i) + (10 + i)$

5. $(4i - 7) + (12 - 4i)$

6. $(a + gi) + (2a + 3gi)$

7. $(7i - 8) - (18 - 2i)$

8. $(3i + 2) - (3i - 2)$

9. $(10 - 5i) - (3 + 2i)$

10. $(2 + 10i) - (6 - 7i)$

Number and Quantity
Set 1: Operations with Complex Numbers

Station 2

Work with your partner to solve each problem. Show all your work. Use the calculator if necessary.

1. $(1 + 3i)(2 + 5i)$

2. $(3 + 7i)(4 + 2i)$

3. $(-1 + 2i)(3 - 2i)$

4. $\left(\dfrac{1}{4} + 2i\right)(10 + i)$

5. $(2i - 3)4i$

6. $(3 - i)(4 + i)$

7. $(8 + 3i)(4 - 5i)$

8. $(10 - 2i^3)(4 + 1)$

9. $(9 + 2i)(9 - 2i)$

Number and Quantity
Set 1: Operations with Complex Numbers

Station 3

Work with your group to identify the conjugate c of complex numbers and then solve each division problem. Show all your work.

1. $\dfrac{3+2i}{5-6i}$

2. $\dfrac{4+3i}{2+i}$

3. $\dfrac{5+2i}{3-2i}$

4. $\dfrac{3+2i}{3-2i}$

5. $\dfrac{7+3i}{7+3i}$

6. $\dfrac{8-3i}{2+i}$

7. $\dfrac{6-2i}{5+3i}$

8. $\dfrac{3-i}{4+2i}$

Number and Quantity
Set 1: Operations with Complex Numbers

Station 4

Work with a group to solve each problem. State your answer in terms of $a + bi$. Show all your work.

1. $8 + \sqrt{-\dfrac{1}{4}}$

2. $\sqrt{-16} + 3$

3. $\sqrt{-9} - 2$

4. $\sqrt{-25}\left(\sqrt{-16} - 4\right)$

5. $\dfrac{4}{3 + \sqrt{-9}}$

6. $\left(2\sqrt{-4}\right)\left(\dfrac{1}{3 - \sqrt{-4}}\right)$

7. $\left(3 + \sqrt{-4}\right)\left(3 - \sqrt{-4}\right)$

8. $\dfrac{1}{4 + \sqrt{-1}} + \dfrac{2 + 2\sqrt{-36}}{3 - \sqrt{-4}}$

Algebra

Set 1: Factoring

Instruction

Goal: To provide opportunities for students to practice factoring quadratic equations

Common Core Standards

Algebra: Seeing Structure in Expressions

Write expressions in equivalent forms to solve problems.

A-SSE.3. Choose and produce an equivalent form of an expression to reveal and explain properties of the quantity represented by the expression.★

   a. Factor a quadratic expression to reveal the zeros of the function it defines.

Student Activities Overview and Answer Key

Station 1

Given equations in the form $y = x^2 + bx + c$, $y = x^2 - bx + c$, $y = x^2 - bx - c$, or $y = x^2 + bx - c$, where b and c are integers, students work in groups to factor by grouping. Students will use algebra tiles as needed.

Answers

1. $(x + 4)(x + 4)$
2. $(x - 7)(x + 5)$
3. $(x - 1)(x + 9)$
4. $(x - 4)(x + 7)$
5. $(x + 2)(x - 1)$
6. $(x + 3)(x + 6)$
7. $(x - 4)(x + 11)$
8. $(x - 10)(x - 3)$
9. $(x - 5)(x + 5)$

Algebra
Set 1: Factoring

Instruction

Station 2

Given equations in the form $y = x^2 + bx + c$, $y = x^2 - bx + c$, $y = x^2 - bx - c$, or $y = x^2 + bx - c$, where b and c are real numbers, students work in pairs to factor by grouping.

Answers

1. $\left(x - \dfrac{1}{2}\right)\left(x - \dfrac{3}{8}\right)$
2. $\left(x + \dfrac{1}{5}\right)(x - 5)$
3. $(x - 8)(x + 3)$
4. $\left(x + \dfrac{1}{3}\right)\left(x + \dfrac{2}{7}\right)$
5. $\left(x - \dfrac{1}{6}\right)(x - 2)$
6. $\left(x + \dfrac{4}{3}\right)(x - 14)$
7. $(x + 0.5)(x + 0.3)$
8. $(x + 0.2)(x - 0.12)$

Station 3

Given equations in the form $y = ax^2 + bx + c$, $y = ax^2 - bx + c$, $y = ax^2 - bx - c$, or $y = ax^2 + bx - c$, where a, b, and c are real numbers, students work in groups to factor by grouping. Students will use algebra tiles as needed.

Answers

1. $4(x + 2)(x - 2)$
2. $3(x - 1)(x - 7)$
3. $5(2x + 3)(x - 6)$
4. $\dfrac{1}{2}(3x - 1)(x - 2)$
5. $\dfrac{2}{5}(x + 4)(x - 10)$

Algebra
Set 1: Factoring

Instruction

6. $7(x+3)(3x+1)$

7. $\dfrac{1}{9}(x-5)(x-2)$

8. $\dfrac{3}{4}(4x-7)(x+2)$

Station 4

Students are given a set of 12 index cards, each inscribed with one of the following expressions: 8, 6, 1/2, 10, 3, (x – 5), (x + 2), (x – 3), (x + 7), (x + 1), (3x + 2), and $\left(x - \dfrac{1}{2}\right)$. They use grouping to factor a series of equations. Each card appears as a factor at least once in this series of equations. Then students combine their cards in pairs to come up with quadratic equations of their own.

Answers

1. $(x-5)(x+1)$

2. $8\left(x - \dfrac{1}{2}\right)(3x+2)$

3. $6(x-3)(3x+2)$

4. $\dfrac{1}{2}(x+7)(x+2)$

5. $10(x+1)\left(x - \dfrac{1}{2}\right)$

6. $3(x-5)(x-3)$

7. Answers will vary. Students should create three quadratic expressions that combine the factors on the cards.

Materials List/Setup

Station 1   algebra tiles

Station 2   none

Station 3   algebra tiles

Station 4   algebra tiles; 12 index cards with the following written on them (one expression per card): 8; 6; 1/2; 10; 3; (x – 5); (x + 2); (x – 3); (x + 7); (x + 1); (3x + 2); $\left(x - \dfrac{1}{2}\right)$

Algebra
Set 1: Factoring

Instruction

Discussion Guide

To support students in reflecting on the activities and to gather some formative information about student learning, use the following prompts to facilitate a class discussion to "debrief" the station activities.

Prompts/Questions

1. What are factors?
2. What is distribution? How does it apply to binomial factors?
3. How do you factor a quadratic equation?
4. If you factor a quadratic equation that is set equal to 0, what points on the equation's graph do the factors represent? Why?

Think, Pair, Share

Have students jot down their own responses to questions, then discuss with a partner (who was not in their station group), and then discuss as a whole class.

Suggested Appropriate Responses

1. Factors are the quantities that are multiplied to produce a product.
2. Distribution is a property of real numbers that allows the multiplication of a term to a sum of terms. With a pair of binomials that are being multiplied together, the distributive property is used twice. Take the first term in the first binomial and multiply it by each term in the second binomial, adding the products. Then take the second term in the first binomial and multiply it by each term in the second binomial, adding all the products together.
3. Rewrite the equation in $y = ax^2 + bx + c$ form, with all x expressions and constants on the same side of the equation. Find the factors of a and the factors of c that combine to create b.
4. They represent the x-intercepts, because those are the points at which $y = 0$.

Possible Misunderstandings/Mistakes

- Incorrectly factoring quadratic expressions
- Incorrectly factoring constants and coefficients
- Not understanding factoring
- Not understanding polynomial factoring
- Not simplifying the equation before factoring
- Making simple arithmetical errors in factoring

Algebra
Set 1: Factoring

Station 1

Work as a group to factor each equation. Use the algebra tiles if you wish. Show all your work.

1. $y = x^2 + 8x + 16$

2. $y = x^2 - 2x - 35$

3. $y = x^2 + 8x - 9$

4. $y = x^2 + 3x - 28$

5. $y = x^2 + x - 2$

6. $y = x^2 + 9x + 18$

7. $y = x^2 + 7x - 44$

8. $y = x^2 - 13x + 30$

9. $y = x^2 - 25$

Algebra
Set 1: Factoring

Station 2

Work in pairs to factor each equation. Show all your work. Check your work by using distribution to find the product of your factors.

1. $y = x^2 - \dfrac{7}{8}x + \dfrac{3}{16}$

2. $y = x^2 - \dfrac{24}{5}x - 1$

3. $y = x^2 - 5x - 24$

4. $y = x^2 + \dfrac{13}{21}x + \dfrac{2}{21}$

5. $y = x^2 - \dfrac{13}{6}x + \dfrac{1}{3}$

6. $y = x^2 - \dfrac{38}{3}x - \dfrac{56}{3}$

7. $y = x^2 + 0.8x + 0.15$

8. $y = x^2 + 0.08x - 0.024$

Algebra
Set 1: Factoring

Station 3

Work as a group to factor each equation. Use the algebra tiles if you wish. Show all your work.

1. $y = 4x^2 - 16$

2. $y = 3x^2 - 24x + 21$

3. $y = 10x^2 - 45x - 90$

4. $y = \dfrac{3}{2}x^2 - \dfrac{7}{2}x + 1$

5. $y = \dfrac{2}{5}x^2 - \dfrac{12}{5}x - \dfrac{80}{5}$

6. $y = 21x^2 + 70x + 21$

7. $y = \dfrac{x^2}{9} - \dfrac{5}{9}x - \dfrac{2}{9}x + \dfrac{10}{9}$

8. $y = 3x^2 + \dfrac{3}{4}x - \dfrac{21}{2}$

Algebra
Set 1: Factoring

Station 4

At this station, you will find algebra tiles and 12 index cards marked with the following expressions:

$$8;\ 6;\ \frac{1}{2};\ 10;\ 3;\ (x-5);\ (x+2);\ (x-3);\ (x+7);\ (x+1);\ (3x+2);\ \left(x-\frac{1}{2}\right)$$

Work as a group to factor each equation below, using the index cards provided. You will also use the factors later in the activity. Use the algebra tiles if you wish. Show all your work.

1. $y = x^2 - 4x - 5$

2. $y = 24x^2 + 4x - 8$

3. $y = 18x^2 - 42x - 36$

4. $y = \frac{1}{2}x^2 + \frac{9}{2}x + 7$

5. $y = 10x^2 + 5x - 5$

6. $y = 3x^2 - 24x + 45$

7. Combine your factor cards to form three different quadratic equations. Write your equations below.

Algebra

Set 2: Solving Quadratics

Instruction

Goal: To provide opportunities for students to solve quadratic equations with real and complex solutions

Common Core Standards

Number and Quantity: The Complex Number System

Use complex numbers in polynomial identities and equations.

N-CN.7. Solve quadratic equations with real coefficients that have complex solutions.

Algebra: Seeing Structure in Expressions

Write expressions in equivalent forms to solve problems.

A-SSE.3. Choose and produce an equivalent form of an expression to reveal and explain properties of the quantity represented by the expression.★

  a. Factor a quadratic expression to reveal the zeros of the function it defines.

Algebra: Reasoning with Equations and Inequalities

Solve equations and inequalities in one variable.

A-REI.4. Solve quadratic equations in one variable.

Solve quadratic equations by inspection (e.g., for $x^2 = 49$), taking square roots, completing the square, the quadratic formula and factoring, as appropriate to the initial form of the equation. Recognize when the quadratic formula gives complex solutions and write them as $a \pm bi$ for real numbers a and b.

Student Activities Overview and Answer Key

Station 1

Students work with a partner to solve the quadratic equations by factoring. Students may use algebra tiles as needed.

Answers

1. $x = -\dfrac{1}{2}, x = 3$
2. $x = \dfrac{1}{3}, x = -2$
3. $x = 7, x = 3$
4. $x = -\dfrac{4}{5}, x = 1$

Algebra
Set 2: Solving Quadratics

Instruction

5. $x = -3, x = 3$
6. $x = 4, x = 2$
7. $x = 2$

Station 2

Students work alone or in groups to solve the quadratic equations by factoring. They should be increasingly comfortable finding real roots independently.

Answers

1. $x = -8, x = \dfrac{1}{9}$
2. $x = 1, x = \dfrac{1}{12}$
3. $x = -2, x = 6$
4. $x = 1, x = -\dfrac{7}{5}$
5. $x = 9$
6. $x = 2, x = 12$
7. $x = -\dfrac{2}{7}, x = \dfrac{5}{2}$

Station 3

Students work in groups to solve quadratic equations using the quadratic formula. They will use a graphing calculator. They begin working with irrational roots.

Answers

1. $x = \dfrac{2 \pm 13.266i}{3}$
2. $x = 3.618, x = 1.382$
3. $x = \dfrac{3 \pm 22.428i}{32}$
4. $x = \dfrac{1 \pm 3.872i}{4}$
5. $x = -9.3589, x = -0.641$

Algebra
Set 2: Solving Quadratics

Instruction

Station 4

Students work in pairs to solve quadratic equations using factoring and the quadratic formula. They will use a graphing calculator. Students may begin to recognize the relationship between the discriminant and the type of roots.

Answers

1. $x = \dfrac{5 \pm 3.873i}{20}$

2. $x = \dfrac{6 \pm 10.392i}{2}$

3. $x = 8$

4. $x = -1.146, x = -7.854$

5. $x = -5, x = -2$

6. $x = -1, x = -2$

7. Students may not have noticed any patterns. Some students may notice the relationship between the discriminant and the type of roots the equation has.

Materials List/Setup

Station 1   algebra tiles
Station 2   none
Station 3   graphing calculator
Station 4   graphing calculator

Algebra
Set 2: Solving Quadratics

Instruction

Discussion Guide

To support students in reflecting on the activities and to gather some formative information about student learning, use the following prompts to facilitate a class discussion to "debrief" the station activities.

Prompts/Questions

1. What is the quadratic formula?
2. What is a real number?
3. What is a complex number?
4. What is the value of i?
5. What is the relationship among factors, x-intercepts, roots, and solutions of a quadratic?

Think, Pair, Share

Have students jot down their own responses to questions, then discuss with a partner (who was not in their station group), and then discuss as a whole class.

Suggested Appropriate Responses

1. If $y = ax^2 + bx + c$, then $x = \dfrac{-b \pm \sqrt{b^2 - 4ac}}{2a}$.
2. A real number is any number that can be expressed as a decimal.
3. A complex number is any number expressed as the sum or difference of a real number and an imaginary number, or $a + bi$.
4. $\sqrt{-1}$
5. A quadratic equation written in factored form shows the factors of the quadratic equation. By taking these factors and setting each of them equal to 0 and solving for x, you are finding the roots or the solutions of the quadratic equation. The roots of the equation are the x-intercepts, or where the graph crosses the x-axis. These are also said to be the solutions. If the equation is in the form "$y =$" or "$f(x) =$", you will be setting that side of the equation equal to 0. This is the equivalent of finding the x-intercepts. Use the zero factor property to set each factor equal to 0 and solve for the variable.

Algebra
Set 2: Solving Quadratics

Instruction

Possible Misunderstandings/Mistakes
- Incorrectly factoring quadratic expressions
- Incorrectly factoring constants and coefficients
- Not understanding factoring
- Not understanding polynomial factoring
- Making simple arithmetical errors in factoring or in applying the quadratic formula
- Incorrectly applying the quadratic formula
- Rounding inaccurately
- Not setting $ax^2 + bx + c$ equal to 0

Algebra
Set 2: Solving Quadratics

Station 1

Work with a partner to solve the quadratic equations by factoring. Use algebra tiles as needed. Show all your work.

1. $2x^2 = 5x + 3$

2. $3x^2 + 5x = 2$

3. $x^2 - 10x = -21$

4. $5x^2 = x + 4$

5. $x^2 - 9 = 0$

6. $2x^2 = 2(6x - 8)$

7. $7x^2 = 28(x - 1)$

Algebra
Set 2: Solving Quadratics

Station 2

Work alone or with your group to solve the quadratic equations by factoring. Show all your work.

1. $-9x^2 + 8 = 71x$

2. $4x^2 = 13x - 1$

3. $\dfrac{x^2}{4} - x = 3$

4. $5x^2 = 7 - 2x$

5. $x^2 = -81 + 18x$

6. $2x - x^2 = 24 - 12x$

7. $31x = 14x^2 - 10$

Algebra
Set 2: Solving Quadratics

Station 3

Work with a partner to solve the problem using the quadratic formula. Use the graphing calculator to estimate square roots to the thousandths. State irrational roots in terms of i.

1. $y = 3x^2 - 4x + 16$

2. $y = x^2 - 5x + 5$

3. $y = 8x^2 - \dfrac{3}{2}x + 4$

4. $y = 2x^2 - x + 2$

5. $y = \dfrac{x^2}{2} + 5x + 3$

Algebra
Set 2: Solving Quadratics

Station 4

Work alone or with your group to solve using either factoring or the quadratic formula. Use the graphing calculator to estimate square roots to the thousandths. State irrational roots in terms of i.

1. $y = 10x^2 - 5x + 1$

2. $y = \dfrac{x^2}{3} - 2x + 12$

3. $y = x^2 - 16x + 64$

4. $y = x^2 + 9x + 9$

5. $y = x^2 + 7x + 10$

6. $y = 3x^2 + 9x + 6$

7. Do you notice any patterns about the value of $\sqrt{b^2 - 4ac}$ and the roots of the equation? If so, what are they?

Algebra

Set 3: Linear Programming

Instruction

Goal: To guide students to an understanding of basic two-variable linear programming, from setting up inequalities through graphing and solving with technology

Common Core Standards

Algebra: Creating Equations★

Create equations that describe numbers or relationships.

A-CED.2. Create equations in two or more variables to represent relationships between quantities; graph equations on coordinate axes with labels and scales.

A-CED.3. Represent constraints by equations or inequalities, and by systems of equations and/or inequalities, and interpret solutions as viable or nonviable options in a modeling context.

Student Activities Overview and Answer Key

Station 1

Working with groups, students analyze word problems to identify variables and set up inequalities and constraints. They use their inequalities and constraints as the basis of graphs. Students find the feasible region of each problem and then find the corner points (vertices) of each feasible region from their graphs. They find the minimum or maximum objective quantity, and determine whether this is a reasonable solution to the problem. Students gain an understanding of which problems can be solved with linear programming in two variables.

Answers

1a. x = production rate of dresses per bolt

y = production rate of skirts per bolt

Maximize $Z = 30x + 20y$

$x \geq 0$

$y \geq 0$

$x \leq 2$

$y \leq 4$

$4x + 2.5y \leq 10$

Algebra
Set 3: Linear Programming

Instruction

1b. x = production rate of dresses per bolt

y = production rate of skirts per bolt

Minimize $W = 10 - (4x + 2.5y)$

$x \geq 1$

$y \geq 0$

$x \leq 2$

$y \leq 4$

$0 \leq W \leq 10$

2a.

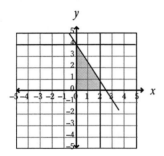

2b.

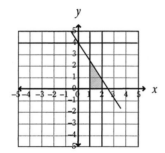

3a. (0, 0), (0, 4), (2, 0), (2, 0.8)

3b. (1, 0), (2, 0), (2, 0.8), (1, 2.4)

4a. Maximum profit from a bolt of fabric: $80, from making 4 skirts and no dresses; the vertex of (2, 0.8) does not yield a reasonable solution, because you can't make 0.8 skirts.

4b. To cut the bolt and produce minimum waste, make one dress and two skirts for total waste of 1 yard. Maximum profit is $70.

5. This is a reasonable solution. Students who have made errors in their work, or who do not fully understand linear programming, may or may not see their solutions as reasonable.

Algebra
Set 3: Linear Programming

Instruction

Station 2

Working with groups, students analyze word problems to identify variables and set up inequalities and constraints. They use their inequalities and constraints as the basis of graphs. Students find the feasible region of each problem and then find the corner points (vertices) of each feasible region from their graphs. They find the minimum or maximum objective quantity, and determine whether this is a reasonable solution to the problem. Students gain an understanding of which problems can be solved with linear programming in two variables.

Answers

1. x = production rate of espressos

   y = production rate of cappuccinos

   Maximize $Z = (1)x + (2.5)y$

   $x \geq 0$

   $y \geq 0$

   $x + y \leq 200$

   $y \leq 125$

2.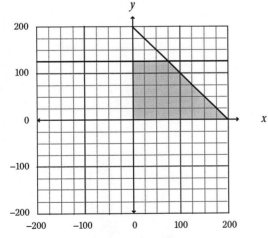

3. (0, 0), (75, 125), (0, 125), (200, 0)

4. Maximum profit: $387.50, with the sale of 75 cups of espresso and 125 cups of cappuccino

5. This is a reasonable solution. Students who have made errors in their work, or who do not fully understand linear programming, may or may not see their solutions as reasonable.

© 2011 Walch Education

Algebra II Station Activities for Common Core State Standards

Algebra
Set 3: Linear Programming

Instruction

Station 3

Working with groups, students analyze word problems to identify variables and set up inequalities and constraints. They use their inequalities and constraints as the basis of graphs. Students find the feasible region of each problem and then find the corner points (vertices) of each feasible region from their graphs. They find the minimum or maximum objective quantity, and determine whether this is a reasonable solution to the problem. Students gain an understanding of which problems can be solved with linear programming in two variables.

Answers

1. x = number of bookcases per week

   y = number of tables per week

   $8x + 5y \leq 40$

   $x \geq 1$

   $y \geq 1$

   $x \leq 4$

   $y \leq 6$

   Maximize $Z = 50x + 40y$

2.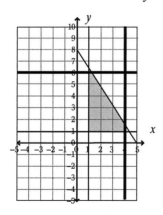

3. (1, 1), (1, 6), (4, 1.6), (4, 1), (1.25, 6)

4. The most profitable way for Jim to use his time is to make 1 bookcase and 6 tables a week. Maximum profit is $290.

5. This is a reasonable solution. The profit is lower at (4, 1). Students may point out that the corner point (1.25, 6) yields a higher profit but does not allow Jim to complete work on the second bookcase in the course of one week; it simply offers him a way to use time left over from making the 1 bookcase and 6 tables.

Algebra
Set 3: Linear Programming

Instruction

Station 4

Working with groups, students analyze a word problem to identify variables and set up inequalities and constraints. Students should recognize that this problem cannot be solved with linear programming. Finally, students provide examples of other situations in which linear programming could help solve problems, and write the inequality and constraints.

Answers

1. Some students may already recognize that the problem cannot be solved with linear programming, and decline to define variables. Those who define variables should recognize four:
   - number of shares in large-cap value fund, x
   - number of shares in overseas mutual fund, y
   - number of shares in municipal bond, m
   - number of shares in technology mutual fund, t
2. cannot be solved
3. cannot be solved; There are four variables.
4. Answers will vary. Students should provide inequalities in two variables and reasonable constraints, including the non-negativity of the two variables.
5. Answers will vary. Some students may suggest using a matrix to solve the system of equations.

Materials List/Setup

Station 1 calculator; colored pens or pencils
Station 2 calculator; colored pens or pencils
Station 3 calculator; colored pens or pencils
Station 4 calculator; colored pens or pencils

Algebra
Set 3: Linear Programming

Instruction

Discussion Guide

To support students in reflecting on the activities and to gather some formative information about student learning, use the following prompts to facilitate a class discussion to "debrief" the station activities.

Prompts/Questions
1. What does "programming" usually mean to you?
2. What is linear programming?
3. What does "optimal" mean?
4. When would you need to use linear programming?

Think, Pair, Share

Have students jot down their own responses to questions, then discuss with a partner (who was not in the same station group), and then discuss as a whole class.

Suggested Appropriate Responses
1. Students may say that they associate the word "programming" with computers.
2. Linear programming is planning that involves sets of linear inequalities.
3. "Optimal" means "best possible."
4. Linear programming may be used in situations in which you need to find the best possible way to use limited resources, or to balance several options.

Possible Misunderstandings/Mistakes
- Incorrectly graphing points
- Incorrectly setting constraints
- Incorrectly assigning variables based on scenario
- Incorrectly writing the function that represents variables' relationship
- Incorrectly using linear programming technology
- Incorrectly applying the formula of a line
- Drawing an inaccurate graph
- Shading the wrong side of a line to create an inaccurate feasible region

Algebra
Set 3: Linear Programming

Instruction

- Making arithmetical errors
- Ignoring one or more constraints
- Incorrectly calculating corner points
- Finding a maximum or minimum that ignores a practical constraint on the quantities involved
- Omitting the constraint that variables are non-negative
- Misunderstanding the scenario that provides the data set
- Students with language or culture barriers may find the scenario setups especially confusing.

NAME: _____

Algebra
Set 3: Linear Programming

Station 1

Work with your group to answer the questions about the following scenario.

> Clara makes and sells clothes. The fabric she uses comes in 10-yard bolts. A dress uses 4 yards of fabric. A skirt uses 2.5 yards. A dress nets a $30 profit; a skirt nets a $20 profit. Clara needs to know the following:

   a. What is the most profitable way to use a bolt of fabric?
   b. If, to satisfy demand, she must make at least one dress from every bolt, what is the least wasteful way for her to use the fabric?

1. Define decision variables and set up an objective function and constraints.

2. Create a graph based on the inequalities and constraints you set up. Label each line in your graph. Shade regions that do not satisfy the inequality.

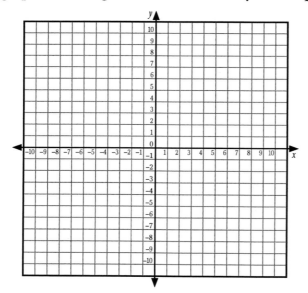

3. Find the corner points of your graph. Show all your work. If you find you need to add an additional constraint to a problem, do so.

continued

Algebra
Set 3: Linear Programming

4. Answer Clara's questions.

   a.

   b.

5. Are your answers reasonable conclusions for the scenario? Why or why not?

Algebra
Set 3: Linear Programming

Station 2

Work with your group to answer the questions about the following scenario.

A coffee shop sells two kinds of espresso-based drinks: shots of espresso and cappuccinos. The profit on an espresso is $1. The profit on a cappuccino is $2.50. Today, the shop has enough beans to make 200 espresso-based drinks. The manager forgot to order cappuccino cups, so the shop can't sell more than 125 cappuccinos. What is the maximum profit the coffee shop can have today?

1. Define decision variables and set up an objective function and constraints.

2. Create a graph based on the inequalities and constraints you set up. Label each line in your graph. Shade regions that do not satisfy the inequality.

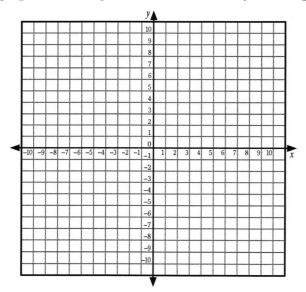

3. Find the corner points of your graph. Show all your work. If you find you need to add an additional constraint to a problem, do so.

4. What is the maximum profit the shop can have today?

5. Is your answer a reasonable conclusion for the scenario? Why or why not?

NAME:

Algebra
Set 3: Linear Programming

Station 3

Work with your group to answer the questions about the following scenario.

> Jim's furniture shop produces bookcases and tables. Bookcases each yield a profit of $50. Tables each yield a $40 profit. Each bookcase takes 8 hours to make. Each table takes 5 hours. Jim works 40 hours a week. Assuming he must make at least 1 bookcase and at least 1 table each week, what is the most profitable way for him to balance his time between bookcases and tables?

1. Define decision variables and set up an objective function and constraints.

2. Create a graph based on the inequalities and constraints you set up. Label each line in your graph. Shade regions that do not satisfy the inequality.

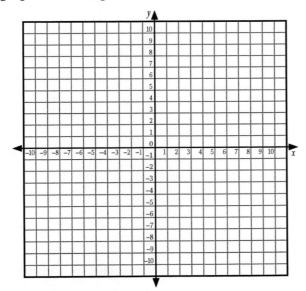

3. Find the corner points of your graph. Show all your work. If you find you need to add an additional constraint to a problem, do so.

4. What is the most profitable way for Jim to balance his time?

5. Are your answers reasonable conclusions for the scenario? Why or why not?

Algebra
Set 3: Linear Programming

Station 4

Work with your group to answer the questions about the following scenario. If you cannot answer a question with linear programming, write "cannot be solved."

> An investor needs to divide $10,000 into a balance of stocks and bonds that will offer the strongest possible return and no more than a 15% chance of total loss. A large-cap value fund offers an average 12% return and carries a 20% chance of loss. An overseas mutual fund offers an average 16% return, with a 30% chance of loss. A municipal bond offers an average 5% return, with a 3% chance of loss. A technology mutual fund offers an average 13% return, with a 12% chance of loss. The technology fund must be bought in shares of $1000 each. What is the most profitable way for the investor to divide her money between these four investments?

1. Define decision variables.

2. Set up an objective function.

3. Can this problem be solved with linear programming? Why or why not?

4. Write a scenario that could be solved with linear programming. Include the decision variables and the objective function.

5. How might you be able to use linear programming with objective functions of more than two variables?

Interpreting Functions

Set 1: Quadratic Transformations in Vertex Form

Instruction

Goal: To provide opportunities for students to analyze the relationship between the equation of a parabola and its graph

Common Core Standards

Algebra: Seeing Structure in Expressions

Write expressions in equivalent forms to solve problems.

A-SSE.3. Choose and produce an equivalent form of an expression to reveal and explain properties of the quantity represented by the expression.★

    b. Complete the square in a quadratic expression to reveal the maximum or minimum value of the function it defines.

Functions: Interpreting Functions

Analyze functions using different representations.

F-IF.7. Graph functions expressed symbolically and show key features of the graph, by hand in simple cases and using technology for more complicated cases.★

    a. Graph linear and quadratic functions and show intercepts, maxima, and minima.

F-IF.8. Write a function defined by an expression in different but equivalent forms to reveal and explain different properties of the function.

    a. Use the process of factoring and completing the square in a quadratic function to show zeros, extreme values, and symmetry of the graph, and interpret these in terms of a context.

Functions: Building Functions

Build new functions from existing functions.

F-BF.3. Identify the effect on the graph of replacing $f(x)$ by $f(x) + k$, $kf(x)$, $f(kx)$, and $f(x + k)$ for specific values of k (both positive and negative); find the value of k given the graphs. Experiment with cases and illustrate an explanation of the effects on the graph using technology. Include recognizing even and odd functions from their graphs and algebraic expressions for them.

Interpreting Functions
Set 1: Quadratic Transformations in Vertex Form

Instruction

Student Activities Overview and Answer Key

Station 1

Given equations in the form $y = x^2 + k$ and $y = (x - h)^2$, where h and k are integers, students graph a series of parabolas, finding the y-intercept and the axis of symmetry. They explore the relationship between the value of h and k and the position of the parabola with respect to the x- and y-axes. Students should also begin to understand the relationship between the equation of the parabola and the axis of symmetry.

Answers

1.

x	y
0	0
1	1
2	4
3	9
–1	1
–2	4
–3	9

2.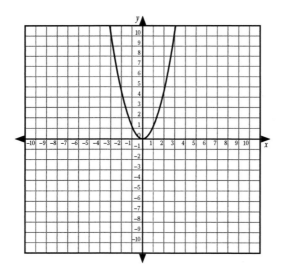

3. (0, 0)

4. (0, 2)

5. $y = x^2 + 2$

6. $y = x^2 - 5$

Interpreting Functions
Set 1: Quadratic Transformations in Vertex Form

Instruction

7.
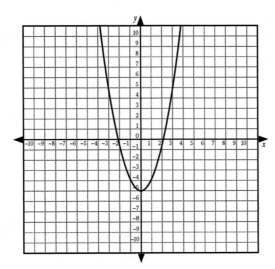

8.
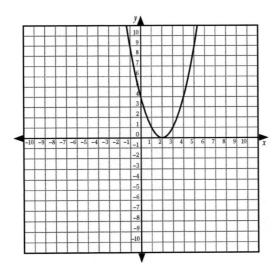

9. $x = 2$

10. $x = -3$

Interpreting Functions
Set 1: Quadratic Transformations in Vertex Form

Instruction

Station 2

Given equations in the form $y = ax^2$, students graph parabolas. Students compare graphs to explore the relationship between the coefficient of x and the width of the parabola.

Answers

1.

x	y
0	0
1	3
2	12
3	27
−1	3
−2	12
−3	27

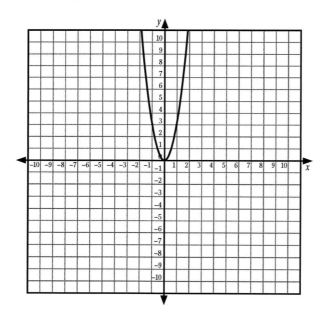

2. (0, 0)

3. $x = 0$

Interpreting Functions
Set 1: Quadratic Transformations in Vertex Form

Instruction

4.

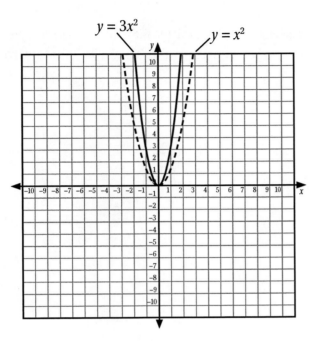

The parabola $3x^2$ is narrower than the parabola x^2.

5–6.

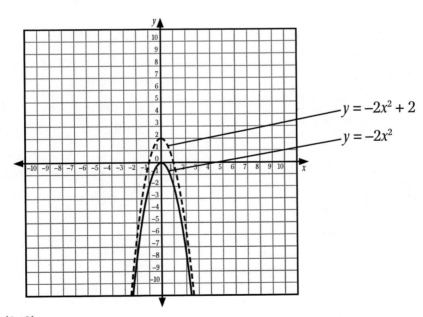

7. (0, 2)

8. The parabola moves vertically.

9. The parabola changes in width.

Interpreting Functions
Set 1: Quadratic Transformations in Vertex Form

Instruction

Station 3

Given equations in the form $y = (x - h)^2 + k$, students graph parabolas. Students find the y-intercept and the axis of symmetry from both the graph and the equation, and begin working towards an understanding of the vertex of a parabola.

Answers

1.

x	y
0	5
1	3
2	5
3	11
4	21
−1	11
−2	21
−3	35

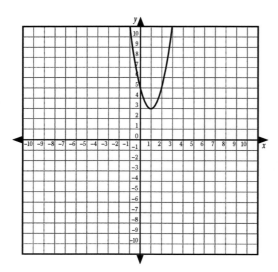

2. $x = 1$

3. $(0, 5)$

Interpreting Functions
Set 1: Quadratic Transformations in Vertex Form

Instruction

4. The parabola would open downward.

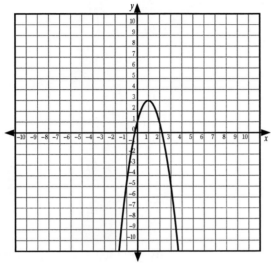

5. (0, 3)

$$y = \frac{1}{2}(x-2)^2 + 1$$

$$y = \frac{1}{2}(0-2)^2 + 1$$

$$y = \frac{1}{2}(4) + 1$$

$$y = 3$$

6. It will be wider, because the higher the coefficient of the x^2 expression, the narrower the parabola. The coefficient of the second x^2 expression is 1, which is higher than ½, the coefficient of the first x^2 expression.

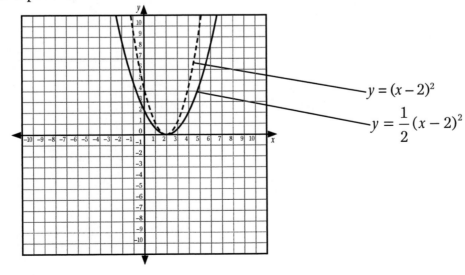

Interpreting Functions
Set 1: Quadratic Transformations in Vertex Form

Instruction

7. (0, 17)

8. $y = 3(x-2)^2 + 10$

   The new y-intercept is (0, 22), so solve for the value of a.

   $y = 3(x-2)^2 + a$
   $22 = 3(0-2)^2 + a$
   $22 = 3(4) + a$
   $22 = 12 + a$
   $10 = a$

Station 4

Students use two methods (completing the square and finding the midpoint of the *x*-intercepts) to convert the equations of parabolas from quadratic form to vertex form. They graph to check their work and to understand the correlation between the different forms and the graph. Students should recognize that a parabola's axis of symmetry always runs through its vertex. They should also understand the relationship between the coordinates of the vertex and the equation in vertex form.

Answers

1.

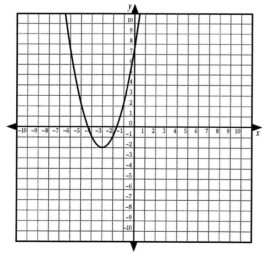

2. $x = -3$

Interpreting Functions
Set 1: Quadratic Transformations in Vertex Form

Instruction

3. $y = x^2 + 6x + 7$
   $y = x^2 + 6x + 7 + 2 - 2$
   $y = (x^2 + 6x + 9) - 2$
   $y = (x + 3)^2 - 2$

4. $(-3, -2)$

5. $(4, 1)$

6. $y = \dfrac{x^2}{2} - 4x + 9$
   $y = \dfrac{x^2}{2} - 4x + 8 + 1$
   $y = \dfrac{1}{2}(x^2 - 8x + 16) + 1$
   $y = \dfrac{1}{2}(x - 4)^2 + 1$

7. $(0, 9)$

8. $x = 4$

9. Yes. The parabola opens out from the vertex. The vertex contains the only y-coordinate that is not repeated in the range.

10. Because the parabola is symmetrical, the axis of symmetry will intersect the midpoint of the line between the x-intercepts. The midpoint is at $(0, 0)$. That means the x-coordinate at the vertex must be 0, because the axis of symmetry intersects the vertex. If $x = 0$, $y = -4$, so the coordinates of the vertex are $(0, -4)$.

Materials List/Setup

Station 1 graph paper
Station 2 colored pencils or pens; graph paper
Station 3 colored pencils or pens; graph paper
Station 4 graph paper

Interpreting Functions
Set 1: Quadratic Transformations in Vertex Form

Instruction

Discussion Guide

To support students in reflecting on the activities and to gather some formative information about student learning, use the following prompts to facilitate a class discussion to "debrief" the station activities.

Prompts/Questions

1. What is a function's axis of symmetry? Does every parabola have one?
2. What is a y-intercept?
3. How do you find the coordinates of a parabola's y-intercept?
4. Compare the equations $y = a(x - h)^2 + k$ and $y = ax^2 - 2axh + ah^2 + k$. Do you think they express the same thing? How could you find out?

Think, Pair, Share

Have students jot down their own responses to questions, then discuss with a partner (who was not in their station group), and then discuss as a whole class.

Suggested Appropriate Responses

1. An axis of symmetry is the line that divides the graph of the function into two symmetrical halves. Every parabola has an axis of symmetry.
2. A y-intercept is the point at which a function crosses the y-axis.
3. Set x equal to 0 and solve the function for y.
4. Students should multiply out the equation in vertex form to find the equation in quadratic form.

Possible Misunderstandings/Mistakes

- Incorrectly calculating the value of y-coordinates from x-coordinates
- Incorrectly graphing parabolas, either from incorrect calculations or from a misunderstanding of graphing itself
- Not understanding the definition of the vertex
- Assuming that the vertex is unrelated to the axis of symmetry
- Incorrectly factoring quadratic equations

Interpreting Functions
Set 1: Quadratic Transformations in Vertex Form

Instruction

- Making simple arithmetical errors in completing the square
- Not understanding the arithmetical manipulations involved in completing the square
- Confusing the *y*-intercept with the *x*-intercept
- Confusing the vertex coordinates h and k
- English language learners may struggle with the questions that ask for written explanations. Encourage these students to write out the numeric operations involved and then describe their work out loud.

NAME: _____

Interpreting Functions
Set 1: Quadratic Transformations in Vertex Form

Station 1

Work as a group to answer the questions. Construct graphs without the aid of a graphing calculator. Show all your work and label the axes of each graph.

1. Given the parabola $y = x^2$, complete the table below with the y coordinates for the following values of x.

x	y
0	
1	
2	
3	
–1	
–2	
–3	

2. Use the coordinates from your table to graph the parabola on graph paper.

3. What are the coordinates for the parabola's y-intercept?

4. Look at the parabola below. What is its y-intercept?

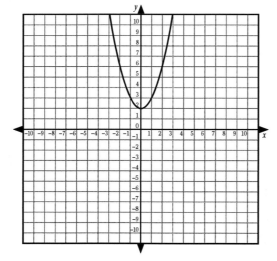

continued

50

Algebra II Station Activities for Common Core State Standards © 2011 Walch Education

Interpreting Functions
Set 1: Quadratic Transformations in Vertex Form

5. What is the equation for this parabola?

6. How would you write the equation for a similar parabola of *y*-intercept (0, –5)?

7. Graph the parabola from problem 6.

8. Graph the parabola $y = (x - 2)^2$.

9. What is the equation for the axis of symmetry of this parabola?

10. Without graphing, predict the equation for the axis of symmetry of the parabola $y = (x + 3)^2$.

Interpreting Functions
Set 1: Quadratic Transformations in Vertex Form

Station 2

Work with your group to explore the relationship between a quadratic function and its graph.

1. Given the equation $y = 3x^2$, complete the table with the values of y and graph the parabola.

x	y
0	
1	
2	
3	
–1	
–2	
–3	

2. What are the coordinates of this parabola's y-intercept?

3. What is the equation of its axis of symmetry?

4. On the graph from problem 1, draw the parabola $y = x^2$ in a contrasting color. In words, compare the two parabolas.

5. Graph the parabola $y = -2x^2$. Complete the table if you need a reference.

x	y
0	
1	
2	
3	
–1	
–2	
–3	

continued

Interpreting Functions
Set 1: Quadratic Transformations in Vertex Form

6. On the same graph, in a contrasting color, graph the parabola $y = -2x^2 + 2$. Label each parabola.

7. What are the coordinates of the y-intercept of $y = -2x^2 + 2$?

8. What happens to the graph of a parabola when you add a numeric constant to its equation, as in problem 6?

9. What happens to the graph of a parabola when the x^2 expression is given a numeric coefficient, as in problems 1 and 5? (*Hint:* Compare the parabola $y = 3x^2$ to $y = x^2$.)

NAME:

Interpreting Functions
Set 1: Quadratic Transformations in Vertex Form

Station 3

Work with your group to answer the following questions.

1. Complete the table for the parabola $y = 2(x - 1)^2 + 3$. Graph the parabola on graph paper.

x	y
0	
1	
2	
3	
4	
−1	
−2	
−3	

2. What is the equation for this parabola's axis of symmetry?

3. What are the coordinates of this parabola's y-intercept?

4. How would this graph change if the parabola's equation changed to $y = -2(x - 1)^2 + 3$? Graph the new parabola to check your answer.

5. What are the coordinates of the y-intercept of the parabola $y = \frac{1}{2}(x - 2)^2 + 1$?

continued

NAME:

Interpreting Functions
Set 1: Quadratic Transformations in Vertex Form

6. Do you think that the graph of $y = \frac{1}{2}(x-2)^2$ will be wider or narrower than the graph of $y = (x-2)^2$? Why? Graph both parabolas, in contrasting colors, to check your answer.

7. Look at the graph below. The equation for this parabola is $y = 3(x-2)^2 + 5$. What is its y-intercept?

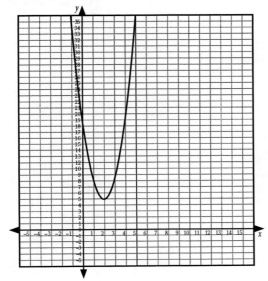

8. How would you write the equation for a similar parabola with a y-intercept 5 units higher? Show your work. Write out an explanation in words if necessary.

Interpreting Functions
Set 1: Quadratic Transformations in Vertex Form

Station 4

Work with your group to answer the following questions.

1. Graph the parabola $y = x^2 + 6x + 7$ on graph paper.

2. Give the equation for its axis of symmetry.

3. *Optional:* Complete the square to give the equation for the parabola in vertex form. Show your work.

4. What are the coordinates of the vertex of this parabola?

5. Look at the graph below. What are the coordinates of the vertex of this parabola?

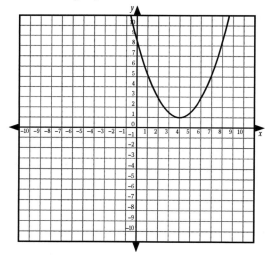

continued

Interpreting Functions
Set 1: Quadratic Transformations in Vertex Form

6. The equation for this parabola is $y = \dfrac{x^2}{2} - 4x + 9$. Find the vertex in order to convert the equation to vertex form. Show your work.

7. What are the coordinates of the y-intercept?

8. What is the equation for the axis of symmetry?

9. Does a parabola's axis of symmetry always run through its vertex? Why or why not?

10. Look at the graph below, which shows the parabola $y = x^2 - 4$. The coordinates of the parabola's x-intercepts are (2, 0) and (–2, 0). How could you use this information to find the coordinates of the parabola's vertex? Explain, showing your work.

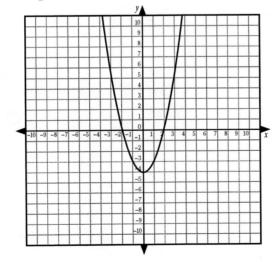

Interpreting Functions

Set 2: Graphing Quadratics

Instruction

Goal: To provide opportunities for students to solve and graph quadratic equations with real solutions, as well as opportunities to analyze graphs and correlate quadratic equations with real-world scenarios

Common Core Standards

Algebra: Reasoning with Equations and Inequalities

Solve equations and inequalities in one variable.

A-REI.4. Solve quadratic equations in one variable.

Solve quadratic equations by inspection (e.g., for $x^2 = 49$), taking square roots, completing the square, the quadratic formula and factoring, as appropriate to the initial form of the equation. Recognize when the quadratic formula gives complex solutions and write them as $a \pm bi$ for real numbers a and b.

Functions: Interpreting Functions

Interpret functions that arise in applications in terms of the context.

F-IF.4. For a function that models a relationship between two quantities, interpret key features of graphs and tables in terms of the quantities, and sketch graphs showing key features given a verbal description of the relationship.★

F-IF.5. Relate the domain of a function to its graph and, where applicable, to the quantitative relationship it describes.★

Analyze functions using different representations.

F-IF.7. Graph functions expressed symbolically and show key features of the graph, by hand in simple cases and using technology for more complicated cases.★

    a. Graph linear and quadratic functions and show intercepts, maxima, and minima.

Student Activities Overview and Answer Key

Station 1

Students work with a partner or a group to solve the quadratic equations using the quadratic formula. Students graph by hand, finding first the vertex $\left(\left(\dfrac{-b}{2a}\right), f\left(\dfrac{-b}{2a}\right)\right)$, then the y-intercept, the point symmetrical to the y-intercept, and other points as needed. Students should recognize the relationship of the vertex to the quadratic equation.

Interpreting Functions
Set 2: Graphing Quadratics

Instruction

Answers

1. $x = \dfrac{-5 \pm \sqrt{10}}{3}$
2. $\left(-\dfrac{5}{3}, -\dfrac{10}{3}\right)$
3. $(0, 5)$
4. $\left(-\dfrac{10}{3}, 5\right)$
5.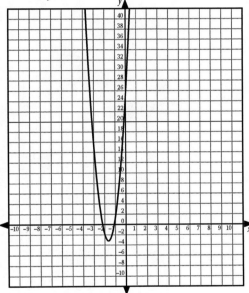
6. The vertex shows the graph's minimum and the location of its axis of symmetry. The other two points provide a rough idea of the curve of the graph.

Station 2

Students work with a partner to solve the quadratic equations using the quadratic formula. Students graph by hand, finding first the vertex $\left(\left(\dfrac{-b}{2a}\right), f\left(\dfrac{-b}{2a}\right)\right)$, then the y-intercept, the point symmetrical to the y-intercept, and other points as needed.

Answers

1. $x = \dfrac{-1 \pm \sqrt{5}}{4}$

Interpreting Functions
Set 2: Graphing Quadratics

Instruction

2. $\left(\dfrac{-1}{4}, \dfrac{-5}{16}\right)$

3. $\left(0, -\dfrac{1}{4}\right)$

4. $\left(-\dfrac{1}{2}, -\dfrac{1}{4}\right)$

5.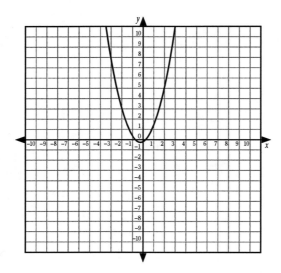

6. no

Station 3

Students work in groups to graph the quadratic model of a stock price using a graphing calculator. Students understand the correlation between the quadratic equation, and the real-world context.

Answers

1. $1.00 or –$9.00 expressed in terms of its break-even value

2. The lowest value will be the *y*-value at the vertex of the curve. Find the vertex at $\left(\left(\dfrac{-b}{2a}\right), f\left(\dfrac{-b}{2a}\right)\right)$. To express the vertex in terms of the break-even value, subtract the price you paid for the stock from the lowest value of the stock, the *y*-value of the vertex. So, $1 – $10 = –$9.

Interpreting Functions
Set 2: Graphing Quadratics

Instruction

3.
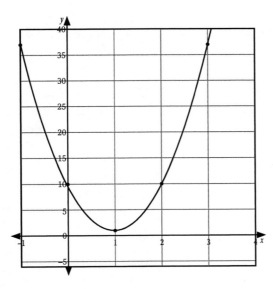

4. 4 P.M., because the vertex falls at the end of the first day

5. To find when the stock price reaches its lowest value again, look at the vertex. This time, look at the *x*-coordinate. The *x*-coordinate represents the time in days. Since the lowest value occurs when the *x*-value is 1, the lowest value occurs at the end of the first day, which is 4 P.M.

6. $37

Station 4

Students work in pairs to solve quadratic equations using the quadratic formula and a graphing calculator. Students understand the correlation between the quadratic formula and real-world scenarios.

Answers

1.
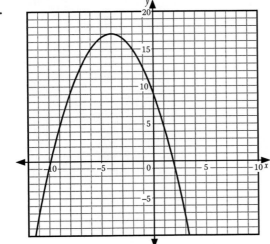

Interpreting Functions
Set 2: Graphing Quadratics

Instruction

2. 17 m

3. The *y*-value of the vertex represents the highest point.

4. 11.7 m

5. The *x*-intercepts are the points where the rocket is launched and lands. The difference between them is the horizontal distance it travels. They can be found by setting $y = 0$ and solving the equation with the quadratic formula. The *x*-intercepts are -9.83095 and 1.83095. So, $1.83095 - (-9.83095) = 11.6619 \approx 11.7$.

Materials List/Setup

Station 1   none
Station 2   none
Station 3   graphing calculator
Station 4   graphing calculator

Interpreting Functions
Set 2: Graphing Quadratics

Instruction

Discussion Guide

To support students in reflecting on the activities and to gather some formative information about student learning, use the following prompts to facilitate a class discussion to "debrief" the station activities.

Prompts/Questions

1. What is the quadratic formula?
2. What is the vertex of a quadratic equation?
3. When is it appropriate to use the quadratic formula as opposed to factoring?
4. What sorts of real-world scenarios might a quadratic equation describe?

Think, Pair, Share

Have students jot down their own responses to questions, then discuss with a partner (who was not in their station group), and then discuss as a whole class.

Suggested Appropriate Responses

1. If $y = ax^2 + bx + c$, $x = \dfrac{-b \pm \sqrt{b^2 - 4ac}}{2a}$.

2. The vertex is the point $\left(\left(\dfrac{-b}{2a} \right), f\left(\dfrac{-b}{2a} \right) \right)$, which represents either the minimum or the maximum y-value of the function as well as its midpoint for all x.

3. It is appropriate to use the quadratic formula when the equation has no rational or real factors.

4. A quadratic equation can describe an area. It can describe simple projectile motion.

Interpreting Functions
Set 2: Graphing Quadratics

Instruction

Possible Misunderstandings/Mistakes

- Incorrectly factoring quadratic expressions
- Incorrectly factoring constants and coefficients
- Not understanding factoring
- Not understanding polynomial factoring
- Making simple arithmetical errors in factoring or in applying the quadratic formula
- Confusing positive and negative signs in finding the vertex
- Not understanding the correlation between the symmetry of the quadratic graph and the y-intercept
- Not understanding the concept of the vertex

Interpreting Functions
Set 2: Graphing Quadratics

Station 1

Work as a group to solve the quadratic equations using the quadratic formula. Graph them by hand, finding first the vertex $\left(\left(\dfrac{-b}{2a}\right), f\left(\dfrac{-b}{2a}\right)\right)$, then the y-intercept, the point symmetrical to the y-intercept, and whatever other points you need. Answer the questions.

1. $3x^2 + 10x + 5 = 0$

2. What are the coordinates of the vertex?

3. What are the coordinates of the y-intercept?

4. Give the coordinates of the point symmetrical to the y-intercept.

5. Plot the graph.

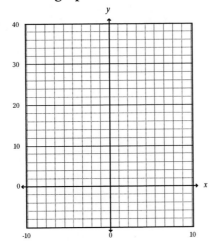

6. In words, explain why finding the vertex, y-intercept, and a point symmetrical to the y-intercept provides an accurate approximation of the graph.

Interpreting Functions
Set 2: Graphing Quadratics

Station 2

Work with a partner to solve the quadratic equations using the quadratic formula. Graph them by hand, finding first the vertex $\left(\left(\dfrac{-b}{2a}\right), f\left(\dfrac{-b}{2a}\right)\right)$, then the y-intercept, the point symmetrical to the y-intercept, and whatever other points you need to show the graph. Answer the questions.

1. $x^2 + \dfrac{x}{2} - \dfrac{1}{4} = 0$

2. What are the coordinates of the vertex?

3. What are the coordinates of the y-intercept?

4. Give the coordinates of the point symmetrical to the y-intercept.

5. Plot the graph.

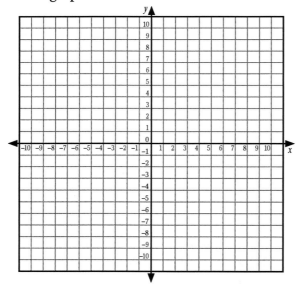

6. Could this equation be solved by factoring?

Interpreting Functions
Set 2: Graphing Quadratics

Station 3

Work with a partner to solve the problem using the quadratic formula. Graph the equation on the graphing calculator. Answer the questions.

1. You buy shares of a company's stock for $8.50 per share. Over two days, the price of your stock drops sharply and then quickly regains its value. The value of the stock can be traced on the curve $y = 8x^2 - 16x + 9$, where each unit on the y-axis is one dollar and each unit on the x-axis represents one day of trading. The line $y = 9$ represents the stock's break-even price. Any y-values above that line represent profit. Any y-values below that line represent loss. What is the lowest value of the stock, expressed in terms of its break-even value?

2. In words, explain how you found the stock's lowest value.

3. Using the domain $0 < x < 2$, graph the stock price for the two days of trading. Use the calculator. Sketch the result here.

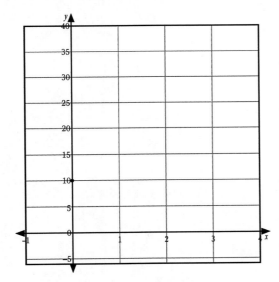

4. Assuming there are 7.5 hours of trading in every day beginning at 9:30 A.M. and ending at 4 P.M., at roughly what time did the stock's price fall hit its lowest value?

5. In words, explain how you found the answer to question 4.

6. If the stock's price continues to follow this curve, what will the stock be worth at the end of the trading day on the third day?

NAME: _____

Interpreting Functions
Set 2: Graphing Quadratics

Station 4

Work with a partner to solve the problem using the quadratic formula. Graph the equation on the graphing calculator. Answer the questions.

1. A model rocket follows the trajectory of $y = -\dfrac{x^2}{2} - 4x + 9$. Graph its flight on the calculator.

    Sketch your result here. Assume that each unit of the graph stands for 1 meter.

    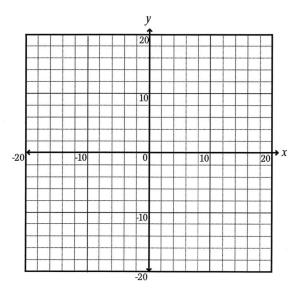

2. Without using the calculator, find the highest point the rocket reached.

3. In words, explain how you found the highest point.

4. How much horizontal distance does the rocket cover? Assume it is launched from ground level and crashes back into the ground.

5. In words, explain how you found the amount of distance.

Interpreting Functions

Set 3: Piecewise Functions

Instruction

Goal: To guide students to an understanding of piecewise functions

Common Core Standards

Functions: Interpreting Functions

Understand the concept of a function and use function notation.

F-IF.2. Use function notation, evaluate functions for inputs in their domains, and interpret statements that use function notation in terms of a context.

Analyze functions using different representations.

F-IF.7. Graph functions expressed symbolically and show key features of the graph, by hand in simple cases and using technology for more complicated cases.★

b. Graph square root, cube root, and piecewise-defined functions, including step functions and absolute value functions.

Student Activities Overview and Answer Key

Station 1

Working with partners, students analyze graphs of different piecewise functions.

Answers

1. $f(x) = \dfrac{4}{3}x - 2$

2. -4

3. $f(x) = 2$

4. $f(x) = \begin{cases} \dfrac{4}{3}x - 2; & \dfrac{3}{2} < x \leq 3 \\ -\sqrt{-2x+3} - 1; & x \leq \dfrac{3}{2} \\ 3; & x > 3 \end{cases}$

5. No, you could not draw the function without having to lift your pencil.

6. $f(x) = 3x - 4$

7. $0 \leq x \leq 2$ or $0 \leq x < 2$

Interpreting Functions
Set 3: Piecewise Functions

Instruction

8. $f(x) = -3x + 20$ for $4 < x \leq 6$ or $4 \leq x \leq 6$, depending on the interval listed in problem 7. The slope is -3.

9. $f(x) = \begin{cases} \dfrac{x^2}{2}; & 0 \leq x < 2 \\ 3x - 4; & 2 \leq x \leq 4 \\ -3x + 20; & 4 < x \leq 6 \end{cases}$

10. $f(2) = 2$. Use $f(x) = \dfrac{x^2}{2}$, since the interval is closed for that function at $x = 2$.

11. Yes, the function is continuous because you could draw the entire function without lifting your pencil. There are no gaps or jumps in the function.

Station 2

Working with groups, students analyze graphs of different piecewise functions.

Answers

1. $f(x) = \begin{cases} 4x; & x > 0 \\ -x^2; & x \leq 0 \end{cases}$

2. Yes, the function is continuous because you could draw the entire function without lifting your pencil. There are no gaps or jumps in the function.

3. -3; use $f(x) = -3$

4. -3; use $f(x) = -3$

5. 1; use $f(x) = \dfrac{x}{2} + 1$

6. 16; use $f(x) = (x - 3)^2$

7. $f(x) = \begin{cases} 2x; & 1 \leq x < 2 \\ 2x + 1; & 2 \leq x < 3 \\ 2x + 2; & 3 \leq x < 4 \\ 2x + 3; & 4 \leq x < 5 \end{cases}$

Station 3

Students work in groups to graph piecewise functions.

Interpreting Functions
Set 3: Piecewise Functions

Instruction

Answers

1.
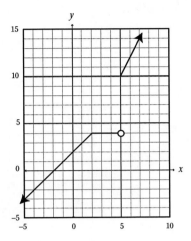

2.
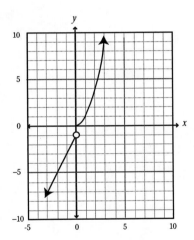

3.
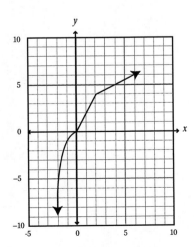

4. Students may mention such applications as the different legs of a triathlon, gradual tax rates, or graduated payment plans.

Interpreting Functions
Set 3: Piecewise Functions

Instruction

Station 4

Students work with partners to graph piecewise functions. Students should recognize the correlation between the slope and the rate of change.

Answers

1.
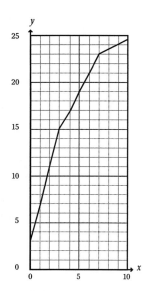

2. The rate of change slows as x rises. The function is gradually flattening out.

3. 11; use $f(x) = 4x + 3$, since that is the function that is used over the interval (0, 3)

4.
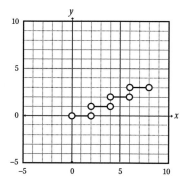

5. $f(x) = 4$

6. The function is undefined at $f(4)$. The intervals are non-inclusive.

7. 3; Since 7.63 falls within the interval (6, 8), use $f(x) = 3$.

8. No, you cannot draw the function without lifting your pencil.

Interpreting Functions
Set 3: Piecewise Functions

Instruction

Materials List/Setup

Station 1 none
Station 2 calculator
Station 3 ruler; colored pens or pencils
Station 4 ruler; colored pens or pencils

Interpreting Functions
Set 3: Piecewise Functions

Instruction

Discussion Guide

To support students in reflecting on the activities and to gather some formative information about student learning, use the following prompts to facilitate a class discussion to "debrief" the station activities.

Prompts/Questions

1. What is a piecewise function?
2. What is a continuous function?
3. What is the difference between a domain and an interval?
4. How might a slope be related to a rate of change?

Think, Pair, Share

Have students jot down their own responses to questions, then discuss with a partner (who was not in their station group), and then discuss as a whole class.

Suggested Appropriate Responses

1. A piecewise function is a function that behaves differently for different values of x.
2. A continuous function is a function whose graph has no gaps or discontinuities.
3. A domain is the set of all x-values for a function. An interval is a set of x-values for which a piecewise function behaves one way.
4. The steeper the slope, the faster the vertical change.

Possible Misunderstandings/Mistakes

- Not understanding the relationship between a function and its graph
- Not understanding the concept of continuous functions
- Not understanding the horizontal-line rule
- Not understanding the difference between $<$ and $\leq$
- Making simple arithmetical errors
- Not understanding the concept of interval or domain
- Not understanding that a slope is a rate of change
- Not understanding (or not applying) the $y = mx + b$ function for a line

NAME: _____

Interpreting Functions
Set 3: Piecewise Functions

Station 1

Work with your partner to analyze each graph.

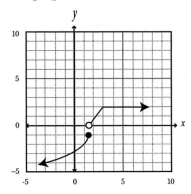

1. Given the points (3, 2) and $\left(\frac{3}{2}, 0\right)$, what is the linear equation of f(x) for $\frac{3}{2} < x \leq 3$?

2. The function $f(x) = -\sqrt{-2x + 3} - 1$ is defined over the interval $x \leq \frac{3}{2}$. Evaluate $f(-3)$.

3. What is the equation of f(x) for x > 3?

4. Define the piecewise function. Write it in function notation. Be sure to include the appropriate intervals.

5. Is the above function continuous? Explain.

continued

Interpreting Functions
Set 3: Piecewise Functions

Use the graph below to answer questions 6–11.

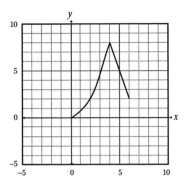

6. Find the linear equation of the function for the interval $2 \leq x < 4$.

7. What is the interval for $f(x) = \dfrac{x^2}{2}$?

8. Define the function and the interval for the remaining piece of the function. What is the slope of the function for this interval?

9. Define the piecewise function. Write it in function notation. Be sure to include the appropriate intervals.

10. Evaluate $f(2)$. Justify your answer.

11. Is the above function continuous? Explain.

Interpreting Functions
Set 3: Piecewise Functions

Station 2

Work with your partner to answer each question. Show all your work. Use the calculator if you need help.

1. Using the points (0, 0), (–1, –1), (–2, –4), (–3, –9), (1, 4), and (2, 8), provide f(x).

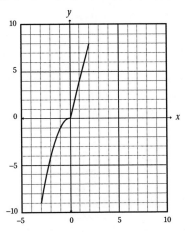

2. Is the above function continuous? Explain.

Use the following piecewise function to evaluate problems 3–6. Justify your solutions.

$$\text{Given } f(x) = \begin{cases} \dfrac{x}{2}+1; & -6 \leq x < 3 \\ -3; & 3 \leq x \leq 5 \\ (x-3)^2; & 5 < x < 9 \end{cases}$$

3. $f(3)$
4. $f(5)$
5. $f(0)$
6. $f(7)$
7. Give the equation for f(x) based on the following graph.

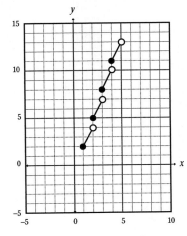

Interpreting Functions
Set 3: Piecewise Functions

Station 3

Work with your group to graph the piecewise functions.

1. $f(x) = \begin{cases} x+2; & x \leq 2 \\ 4; & 2 < x < 5 \\ 2x; & 5 \leq x \end{cases}$

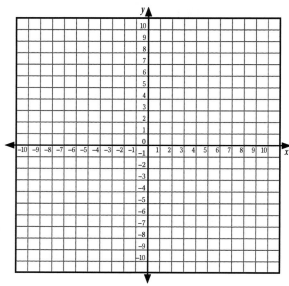

2. $f(x) = \begin{cases} 2x-1; & x < 0 \\ x^2; & 0 \leq x \end{cases}$

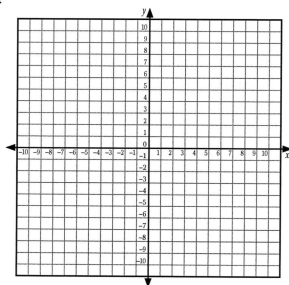

continued

Interpreting Functions
Set 3: Piecewise Functions

3. $f(x) = \begin{cases} x^3; & x \leq 0 \\ 2x; & 0 < x < 2 \\ \dfrac{x}{2} + 3; & 2 \leq x \end{cases}$

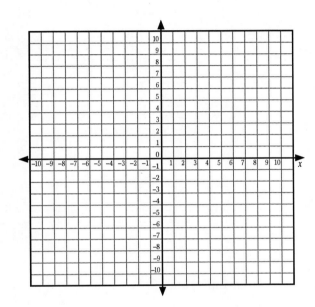

4. What might be some examples of real-life scenarios that involve piecewise functions?

Interpreting Functions
Set 3: Piecewise Functions

Station 4

Work with your group to graph each function and answer the questions. Show all your work.

1. $f(x) = \begin{cases} 4x+3; & 0 \leq x < 3 \\ 2x+9; & 3 < x < 7 \\ \dfrac{x}{2} + \dfrac{39}{2}; & 7 \leq x \leq 10 \end{cases}$

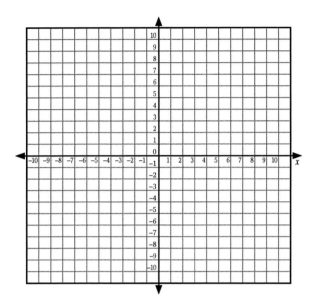

2. What is happening to the function's rate of change as the value of x goes up?

3. Evaluate the function at $f(2)$. Justify your answer.

4. Graph the following function:

$f(x) = \begin{cases} 0; & 0 < x < 2 \\ 1; & 2 < x < 4 \\ 2; & 4 < x < 6 \\ 3; & 6 < x < 8 \end{cases}$

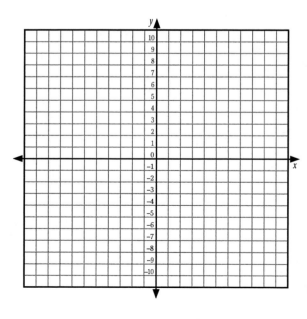

continued

Interpreting Functions
Set 3: Piecewise Functions

5. What can you predict will be the value of $f(x)$ for $8 < x < 10$?

6. Evaluate the function at $f(4)$. Justify your answer.

7. Evaluate the function at $f(7.63)$. Justify your answer.

8. Is this function continuous? Explain.

Interpreting Functions

Set 4: Absolute Value Equations and Inequalities

Instruction

Goal: To guide students to an understanding of solving absolute value equations and inequalities and interpreting the solutions

Common Core Standards

Functions: Interpreting Functions

Analyze functions using different representations.

F-IF.7. Graph functions expressed symbolically and show key features of the graph, by hand in simple cases and using technology for more complicated cases.★

  b. Graph square root, cube root, and piecewise-defined functions, including step functions and absolute value functions.

Student Activities Overview and Answer Key

Station 1

Working with groups, students solve absolute value equations by graphing. Students should recognize the correlation between the solutions and the horizontal line that intersects the absolute value graph.

Answers

1. $x = 1, -1$
2. $x = \dfrac{1}{2}, \dfrac{-3}{2}$
3. $x = 7, 3$
4. $x = -3, 3$
5. no solutions

Station 2

Working with partners, students solve absolute value equations.

Answers

1. $x = 5, -1$
2. $x = -3, -5$
3. $x = 2, -22$

Interpreting Functions
Set 4: Absolute Value Equations and Inequalities

Instruction

4. $x = 14, 2$
5. $x = 8, 10$
6. $x = 3, -\dfrac{7}{2}$
7. Answers will vary. Students should give examples of when absolute value might be applicable.

Station 3

Working with groups, students solve absolute value equations and inequalities.

Answers

1. $x = 3, 0$
2. $x = 8, -8$
3. $x = 3, -\dfrac{11}{3}$
4. $x > 0, x < -\dfrac{4}{3}$
5. $1 < x < 10$
6. $-3 < x < 3$
7. $x = -3, x = -2$

Station 4

Working with groups, students solve absolute value equations and inequalities. They use the graphing calculator as needed.

Answers

1. $x = -4.236, 0.236$
2. $x = 3, 5, 9.568, -1.568$
3. $x = \dfrac{1}{2}, -\dfrac{1}{2}$
4. $x = 3$
5. $x = 7.531, -0.531, 3, 4$

Interpreting Functions
Set 4: Absolute Value Equations and Inequalities

Instruction

Materials List/Setup

Station 1 colored pens or pencils
Station 2 none
Station 3 none
Station 4 graphing calculator

Interpreting Functions
Set 4: Absolute Value Equations and Inequalities

Instruction

Discussion Guide

To support students in reflecting on the activities and to gather some formative information about student learning, use the following prompts to facilitate a class discussion to "debrief" the station activities.

Prompts/Questions
1. What is absolute value?
2. What is an inequality?
3. What does it mean when two graphs intersect?
4. How many solutions does an absolute value inequality usually have?

Think, Pair, Share

Have students jot down their own responses to questions, then discuss with a partner (who was not in their station group), and then discuss as a whole class.

Suggested Appropriate Responses
1. Absolute value means numeric value without regard to whether a quantity is positive or negative.
2. An inequality is a relationship between two different or unequal amounts.
3. When two graphs intersect, the same point falls into both functions.
4. An absolute value inequality usually has four solutions.

Possible Misunderstandings/Mistakes
- Not understanding that a minus sign can mean "opposite" rather than "negative"
- Incorrectly manipulating numbers, variables, or exponents
- Incorrectly factoring
- Incorrectly applying the quadratic equation
- Not understanding the meaning of a < or > sign in relation to absolute value
- Not understanding the graphical meaning of the intersection of an absolute-value graph with the horizontal line of its solution
- Not understanding the concept of absolute value
- Assuming that every absolute-value graph must be linear

Interpreting Functions
Set 4: Absolute Value Equations and Inequalities

Station 1

Work with your partner to solve each equation. Show your work.

1. $|x| = 1$

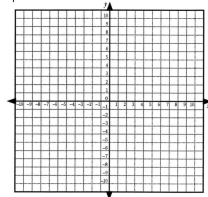

2. $|2x + 1| = 2$

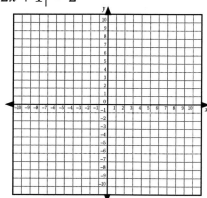

3. $|x - 5| = 2$

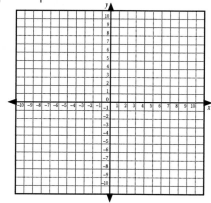

continued

Interpreting Functions
Set 4: Absolute Value Equations and Inequalities

4. $|x^2| = 9$

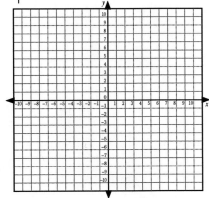

5. $|x^2 + 2| = -1$

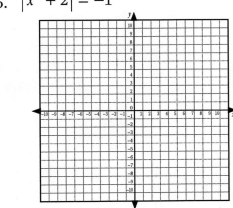

NAME:

Interpreting Functions
Set 4: Absolute Value Equations and Inequalities

Station 2

Work with your partner to solve each equation. Show all your work.

1. $|x - 2| = 3$

2. $|x + 4| = 1$

3. $|x + 10| = 12$

4. $|x - 8| = 6$

5. $|9 - x| = 1$

6. $|4x + 1| - 5 = 8$

7. In what real-life scenarios might absolute value be applicable?

Interpreting Functions
Set 4: Absolute Value Equations and Inequalities

Station 3

Work with your group to solve each equation or inequality. Show all your work.

1. $|2x - 3| + 7 = 10$

2. $\left|\dfrac{x}{2}\right| = 4$

3. $|3x + 1| - 5 = 5$

4. $|3x + 2| > 2$

5. $|11 - 2x| < 9$

6. $|5x| < 15$

7. $|x^2 + 5x + 6| = 0$

Interpreting Functions
Set 4: Absolute Value Equations and Inequalities

Station 4

Work with your group to solve each equation. Use the graphing calculator as needed.

1. $\left|x^2 + 4x + 2\right| = 3$

2. $\left|x^2 - 8x\right| = 15$

3. $\left|4x^2 - 1\right| = 0$

4. $\left|(x-3)^2\right| = 0$

5. $\left|x^2 - 7x + 4\right| = 8$

Interpreting Functions

Set 5: Exponential Functions

Instruction

Goal: To guide students to an understanding of the properties and graphs of exponential functions

Common Core Standards

Functions: Interpreting Functions

Understand the concept of a function and use function notation.

- **F-IF.2.** Use function notation, evaluate functions for inputs in their domains, and interpret statements that use function notation in terms of a context.

Analyze functions using different representations.

- **F-IF.7.** Graph functions expressed symbolically and show key features of the graph, by hand in simple cases and using technology for more complicated cases.★
  - e. Graph exponential and logarithmic functions, showing intercepts and end behavior, and trigonometric functions, showing period, midline, and amplitude.

Student Activities Overview and Answer Key

Station 1

Working with partners, students solve simple problems that explore the properties of exponents.

Answers

1. $f(4) = 81$
2. $f(2) = 16$
   $f(3) = 512$
3. $f(7) = 128$
   $f\left(\dfrac{1}{2}\right) = \sqrt{2}$
   $f(0) = 1$
4. $f(10) = 1$
   $f(2) = 1$
5. $f(2) = 100$
   $f(4) = 10{,}000$
   $f(6) = 1{,}000{,}000$
6. $f(g(x)) = a$

Interpreting Functions
Set 5: Exponential Functions

Instruction

7. $f(1) = 5$
   $f(3) = 125$
   $f(0) = 1$

8. $f(4) = \dfrac{1}{16}$
   $f(2) = \dfrac{1}{4}$
   $f(5) = \dfrac{1}{32}$
   $f(-2) = 4$

9. $f(4) = 4$
   $f(0) = \dfrac{1}{4}$
   $f(3) = 2$
   $f(-2) = \dfrac{1}{16}$

10. $f(1) = 8$
    $f(2) = 64$

Station 2
Working with groups, students determine properties of the graphs of exponential functions.

Answers
1. a. (0, 1)
   b. $y > 0$
   c. none
   d. $y = 0$

Interpreting Functions
Set 5: Exponential Functions

Instruction

2. a. all x
   b. $y > -1$
   c. $y = -1$
   d. (0, 0)
3. a. all x
   b. $y > 3$
   c. $y = 3$
   d. none
4. a. all x
   b. $y > 0$
   c. $y = 0$
   d. none

Station 3

Working with groups, students determine the end behavior of exponential functions. Students use their observations to determine based on the formula whether a formula represents exponential growth or decay.

Answers

1. a. approaches 0
   b. grows without bound
2. a. grows without bound
   b. approaches 0
3. a. decreases without bound
   b. approaches 0
4. Exponential decay; the function approaches 0 as x gets infinitely large.
5. Exponential growth; the function approaches infinity or grows without bound as x gets infinitely large.

Interpreting Functions
Set 5: Exponential Functions

Instruction

Station 4

Student pairs graph exponential functions, checking their work with a graphing calculator.

Answers

1.
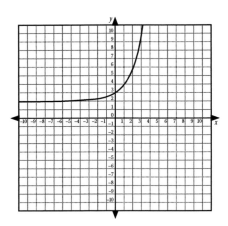

2.
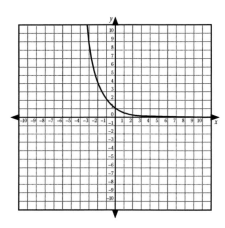

3.
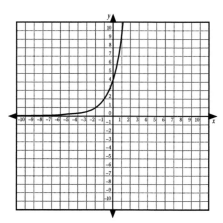

4.
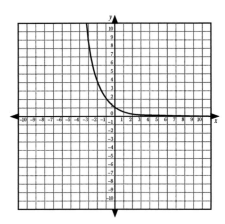

5.
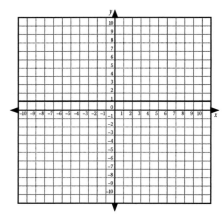

6.
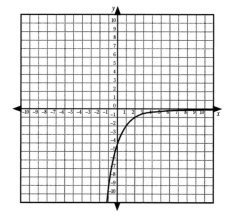

Interpreting Functions
Set 5: Exponential Functions

Materials List/Setup

Station 1 none

Station 2 none

Station 3 none

Station 4 graphing calculator; colored pens or pencils; graph paper

Interpreting Functions
Set 5: Exponential Functions

Instruction

Discussion Guide

To support students in reflecting on the activities and to gather some formative information about student learning, use the following prompts to facilitate a class discussion to "debrief" the station activities.

Prompts/Questions

1. What is an exponent?
2. What is an exponential function?
3. How is an exponential function different from a power function?
4. How do you think we can solve an exponential function?

Think, Pair, Share

Have students jot down their own responses to questions, then discuss with a partner (who was not in their station group), and then discuss as a whole class.

Suggested Appropriate Responses

1. An exponent is a number that tells the power of another number, the base.
2. An exponential function is a function in which the variable occupies the exponent position.
3. In a power function, the variable is the base; in an exponential function, it is the exponent.
4. Give each side an equal base and then set the variables equal to each other. (Students may not yet know how to solve an exponential function. Guide them to a recognition that using the function's inverse may be one way to get the variable into solvable base territory.)

Possible Misunderstandings/Mistakes

- Incorrectly manipulating numbers, variables, or exponents
- Not understanding the laws of exponents
- Assuming that all functions have zeros
- Assuming that an exponential function has a vertical asymptote as it tends toward unbounded growth
- Not understanding the difference between growth and decay
- Incorrectly calculating squares, cubes, etc., of integers between 1 and 10
- Confusing a negative exponent with a fractional exponent
- Not understanding the notation $x \to \infty$

Interpreting Functions
Set 5: Exponential Functions

Station 1

Work with your partner to solve each problem. Show your work.

1. $f(x) = 3^x$
   $f(4) =$

2. $f(x) = 2^{x^2}$
   $f(2) =$
   $f(3) =$

3. $f(x) = 2^x$
   $f(7) =$
   $f\left(\dfrac{1}{2}\right) =$
   $f(0) =$

4. $f(x) = 1^x$
   $f(10) =$
   $f(2) =$

5. $f(x) = 10^x$
   $f(2) =$
   $f(4) =$
   $f(6) =$

6. $f(x) = a^x$
   $g(x) = a^{\frac{1}{x}}$
   $f(g(x)) =$

7. $f(x) = 5^x$
   $f(1) =$
   $f(3) =$
   $f(0) =$

8. $f(x) = 2^{-x}$
   $f(4) =$
   $f(2) =$
   $f(5) =$
   $f(-2) =$

9. $f(x) = 2^{x-2}$
   $f(4) =$
   $f(0) =$
   $f(3) =$
   $f(-2) =$

10. $f(x) = 2^{3x}$
    $f(1) =$
    $f(2) =$

Interpreting Functions
Set 5: Exponential Functions

Station 2

Work with your group to answer each question. Show all your work.

1. $y = b^x$

    a. Where is the y-intercept?

    b. What is the range?

    c. Does the function have any zeros? If so, where are they?

    d. Does the function have any asymptotes? If so, where?

2. $y = 2^x - 1$

    a. What is the domain?

    b. What is the range?

    c. Does the function have any asymptotes? If so, where?

    d. Does the function have any zeros? If so, where are they?

3. $y = a^{-x} + 3$

    a. What is the domain?

    b. What is the range?

    c. Does the function have any asymptotes? If so, where?

    d. Does the function have any zeros? If so, where are they?

4. $y = \left(2^x\right)^5$

    a. What is the domain?

    b. What is the range?

    c. Does the function have any asymptotes? If so, where?

    d. Does the function have any zeros? If so, where are they?

Interpreting Functions
Set 5: Exponential Functions

Station 3

Work with a group to answer each question. Show all your work.

1. $y = 0.5^x$

   a. What is the end behavior as $x \to \infty$?

   b. What is the end behavior as $x \to -\infty$?

2. $y = 3^x$

   a. What is the end behavior as $x \to \infty$?

   b. What is the end behavior as $x \to -\infty$?

3. $y = -2(2^x)$

   a. What is the end behavior as $x \to \infty$?

   b. What is the end behavior as $x \to -\infty$?

4. $y = \left(\dfrac{1}{4}\right)^x$

   Does this function represent exponential growth or decay? Explain.

5. $y = 4^x$

   Does this function represent exponential growth or decay? Explain.

Interpreting Functions
Set 5: Exponential Functions

Station 4

Work with a partner to graph each equation on graph paper. Check your work with a graphing calculator.

1. $y = 2^x + 2$

2. $y = \left(\dfrac{1}{2}\right)^x$

3. $y = 2^{x+2}$

4. $y = 2^{-x}$

5. $y = 1^x$

6. $y = -4\left(\dfrac{1}{2}\right)^x$

Interpreting Functions

Set 6: Solving Exponential Equations and Inequalities

Instruction

Goal: To guide students to an ability to solve and graph exponential functions without using logarithms or e

Common Core Standards

Functions: Interpreting Functions

Understand the concept of a function and use function notation.

F-IF.2. Use function notation, evaluate functions for inputs in their domains, and interpret statements that use function notation in terms of a context.

Analyze functions using different representations.

F-IF.7. Graph functions expressed symbolically and show key features of the graph, by hand in simple cases and using technology for more complicated cases.★

e. Graph exponential and logarithmic functions, showing intercepts and end behavior, and trigonometric functions, showing period, midline, and amplitude.

Student Activities Overview and Answer Key

Station 1

Working with partners, students solve simple exponential equations and inequalities by giving each side an equal base.

Answers

1. $x = 3$
2. $x = 4$
3. $x = -2$
4. $x = -2$
5. $x < 2$
6. $x = 3$
7. $x > 4$
8. $x = \dfrac{1}{4}$

Interpreting Functions
Set 6: Solving Exponential Equations and Inequalities

Instruction

Station 2

Working with groups, students determine the *y*-intercepts and solutions to exponential functions using their graphs. Then, students are given a pair of points and asked to determine the exponential function that passes through those points.

Answers

1.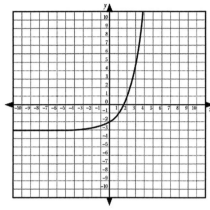

   y-intercept: (0, –2)

   $1 < x < 2$

2.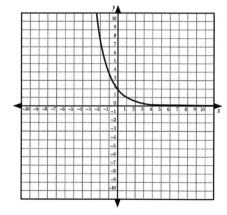

   y-intercept: (0, 2)

   no solution

Interpreting Functions
Set 6: Solving Exponential Equations and Inequalities

Instruction

3.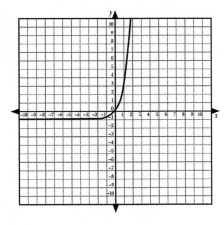

solution at (0, 0)

4.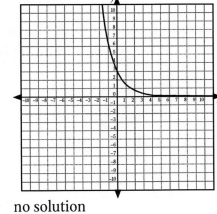

no solution

5.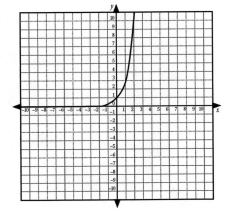

$y = 3^x$

Interpreting Functions
Set 6: Solving Exponential Equations and Inequalities

Instruction

Station 3

Working with groups, students use calculators to evaluate and graph exponential functions.

Answers

1. $f(4) = 0.31640625$

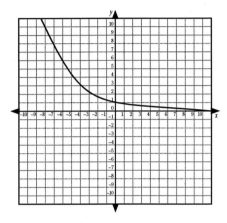

2. $f(6) = \dfrac{1}{729}$

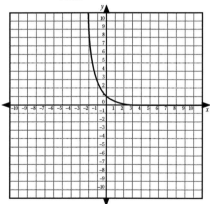

3. $f(5) = 1.01024$

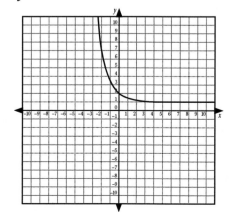

Interpreting Functions
Set 6: Solving Exponential Equations and Inequalities

Instruction

4.
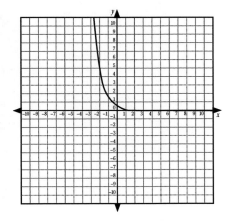
no solutions

5.
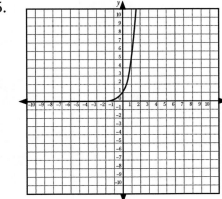
no solutions

6.
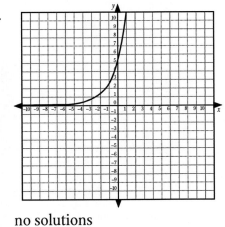
no solutions

Interpreting Functions
Set 6: Solving Exponential Equations and Inequalities

Instruction

7.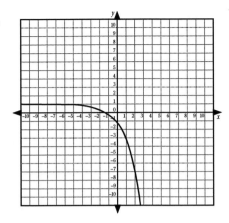

   solution at (–1, 0)

8. The graph must cross the *x*-axis, so the equation must include an addition or subtraction operation in addition to the exponential operation.

Station 4

Working with groups, students use exponential functions to calculate compound interest according to the formula $A = P\left(1 + \dfrac{r}{m}\right)^{tm}$.

Answers

1. $A = 2500\left(1 + \dfrac{0.035}{4}\right)^{4 \cdot 4}$
   $A = \$2873.93$

2. $A = 5000\left(1 + \dfrac{0.027}{12}\right)^{2 \cdot 12}$
   $A = \$5277.10$

3. $A = 500\left(1 + \dfrac{0.04}{6}\right)^{5 \cdot 6}$
   $A = \$610.30$

4. $A = 6000\left(1 + \dfrac{0.045}{12}\right)^{3 \cdot 12}$
   $A = \$6865.49$

Interpreting Functions
Set 6: Solving Exponential Equations and Inequalities

Instruction

5. $A = 1000\left(1+\dfrac{0.05}{4}\right)^{2\cdot 4} \approx 1104.49$

    $A = 1000\left(1+\dfrac{0.046}{12}\right)^{2\cdot 12} \approx 1096.17$

    The account with 5% interest has the better yield since that account will yield approximately $1104.49 and the account with the 4.6% interest rate will yield approximately $1096.17.

Materials List/Setup

Station 1   none
Station 2   colored pens or pencils
Station 3   graphing calculator; colored pens or pencils
Station 4   calculator

Interpreting Functions
Set 6: Solving Exponential Equations and Inequalities

Instruction

Discussion Guide

To support students in reflecting on the activities and to gather some formative information about student learning, use the following prompts to facilitate a class discussion to "debrief" the station activities.

Prompts/Questions

1. What is an exponential function?
2. When does a function have a solution?
3. What is compound interest?
4. Why can it be difficult to estimate compound interest?

Think, Pair, Share

Have students jot down their own responses to questions, then discuss with a partner (who was not in their station group), and then discuss as a whole class.

Suggested Appropriate Responses

1. An exponential function is a function in which the variable occupies the exponent position.
2. A function has a solution when its graph crosses the x-axis.
3. Compound interest is interest that accumulates according to the total (principal plus interest) already in the account, not just according to the principal.
4. The amount on which the percentage is based keeps changing.

Possible Misunderstandings/Mistakes

- Incorrectly manipulating numbers, variables, or exponents
- Not understanding the laws of exponents
- Assuming that all functions have zeros
- Not understanding the difference between growth and decay
- Incorrectly calculating squares, cubes, etc., of integers between 1 and 10
- Confusing a negative exponent with a fractional exponent
- Incorrectly using the exponent function of a calculator
- Incorrectly applying the formula of compound interest
- Not understanding the relationship between an exponential function and its graph

Interpreting Functions
Set 6: Solving Exponential Equations and Inequalities

Station 1

Work with your partner to solve for *x*. Show your work.

1. $8 = 2^x$

2. $81 = 3^x$

3. $\dfrac{1}{25} = 5^x$

4. $4 = \left(\dfrac{1}{2}\right)^x$

5. $100 > 10^x$

6. $223 = 6^x + 7$

7. $0.0625 > 0.5^x$

8. $7 = 2401^x$

Interpreting Functions
Set 6: Solving Exponential Equations and Inequalities

Station 2

Work with your group to answer each question. Remember that the rate of growth r can be found with the formula $y = a(1 + r)^x$.

1. Graph $y = 2^x - 3$.

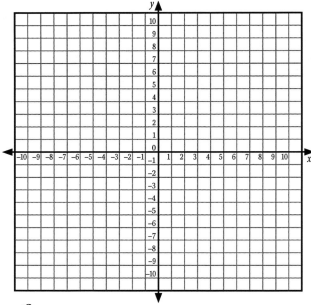

   Where is the y-intercept?

   Does the equation have a solution? If so, estimate where it is.

2. Graph $y = \left(\dfrac{1}{2}\right)^{x-1}$.

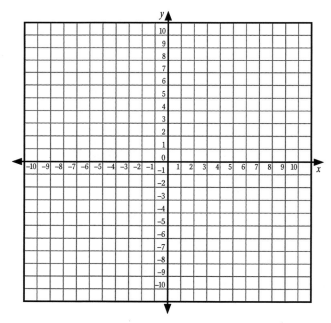

   Where is the y-intercept?

   Does the equation have a solution? If so, estimate where it is.

continued

Interpreting Functions
Set 6: Solving Exponential Equations and Inequalities

3. Graph $y = 4^x - 1$.

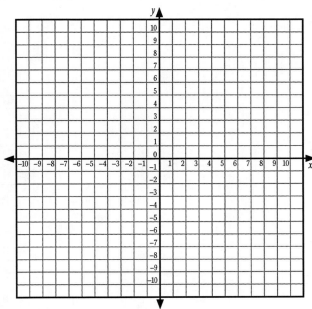

Does the equation have a solution? If so, estimate where it is.

4. Graph $y = 3\left(\dfrac{1}{2}\right)^x$.

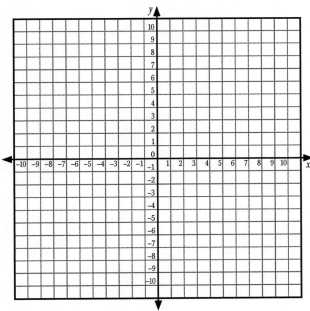

Does the equation have a solution? If so, estimate where it is.

continued

Interpreting Functions
Set 6: Solving Exponential Equations and Inequalities

5. An exponential function passes through the points (0, 1) and (2, 9). What is the function? Graph your answer.

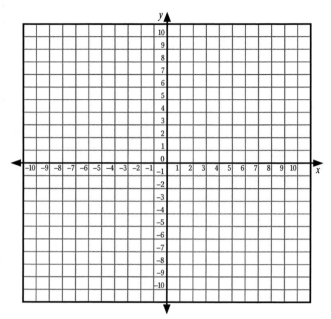

Interpreting Functions
Set 6: Solving Exponential Equations and Inequalities

Station 3

Using a calculator, work with your group to solve each problem. Sketch the graphs.

1. $f(x) = 0.75^x$

   $f(4) =$

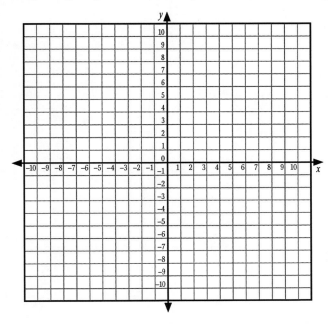

2. $f(x) = 3^{-x}$

   $f(6) =$

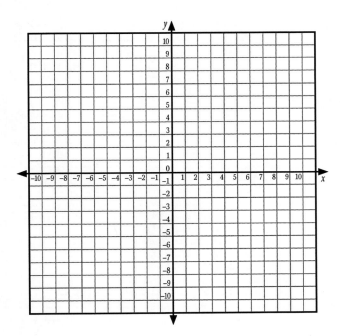

continued

Interpreting Functions
Set 6: Solving Exponential Equations and Inequalities

3. $f(x) = 0.4^x + 1$

   $f(5) =$

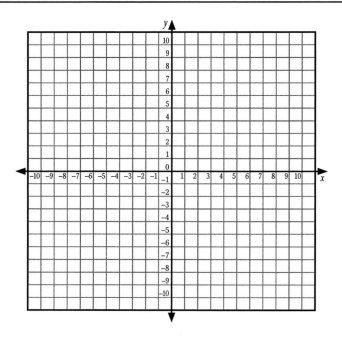

4. Graph $y = \dfrac{1}{2}\left(\dfrac{1}{3}\right)^x$. If there is a solution, what is it?

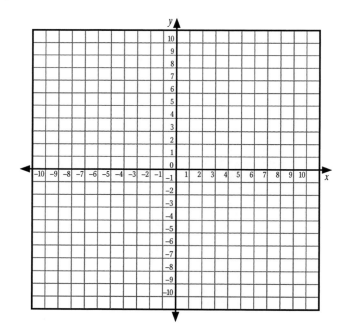

continued

Interpreting Functions
Set 6: Solving Exponential Equations and Inequalities

5. Graph $y = 4^x$. If there is a solution, what is it?

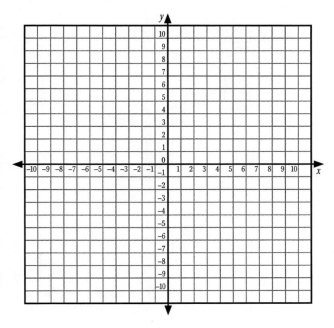

6. Graph $y = 5(2^x)$. If there is a solution, what is it?

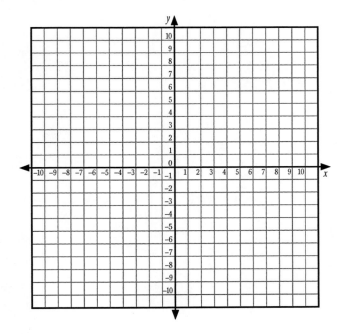

Interpreting Functions
Set 6: Solving Exponential Equations and Inequalities

7. Graph $y = -2(2^x) + 1$. If there is a solution, what is it?

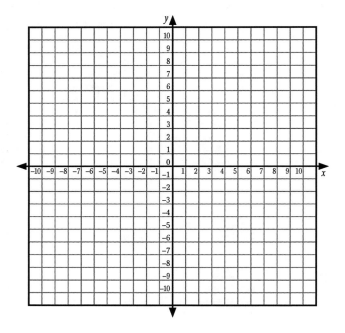

8. For an exponential function to have a solution, what must be true of the equation?

Interpreting Functions
Set 6: Solving Exponential Equations and Inequalities

Station 4

The formula for compound interest is $A = P\left(1 + \dfrac{r}{m}\right)^{tm}$, where A is the final total (principal plus interest), P is the initial amount (principal), r is the interest rate, t is the amount of time in years, and m is the number of times the interest compounds per year. Work with your group to set up and then calculate each equation. Round answers to the nearest penny.

1. An account with an initial balance of $2500 has interest of 3.5% that compounds quarterly over four years. What is the balance at the end of the fourth year?

2. An account with an initial balance of $5000 has interest of 2.7% that compounds monthly over two years. What is the balance at the end of the second year?

3. An account with an initial balance of $500 has interest of 4% that compounds every other month over five years. What is the balance at the end of the fifth year?

4. An account with an initial balance of $6000 has interest of 4.5% that compounds monthly over three years. What is the balance at the end of the third year?

5. If you have $1000 to invest for two years, which account has the better yield: an account that compounds quarterly at 5%, or one that compounds monthly at 4.6%?

Interpreting Functions

Set 7: Polynomial Functions

Instruction

Goal: To guide students to facility with evaluating, factoring, and graphing polynomials, both by hand and with technology

Common Core Standards

Number and Quantity: The Complex Number System

Use complex numbers in polynomial identities and equations.

N-CN.9. (+) Know the Fundamental Theorem of Algebra; show that it is true for quadratic polynomials.

Algebra: Seeing Structure in Expressions

Interpret the structure of expressions.

A-SSE.2. Use the structure of an expression to identify ways to rewrite it.

Algebra: Arithmetic with Polynomials and Rational Expressions

Understand the relationship between zeros and factors of polynomials.

A-APR.2. Know and apply the Remainder Theorem: For a polynomial $p(x)$ and a number a, the remainder on division by $x - a$ is $p(a)$, so $p(a) = 0$ if and only if $(x - a)$ is a factor of $p(x)$.

A-APR.3. Identify zeros of polynomials when suitable factorizations are available, and use the zeros to construct a rough graph of the function defined by the polynomial.

Functions: Interpreting Functions

Understand the concept of a function and use function notation.

F-IF.2. Use function notation, evaluate functions for inputs in their domains, and interpret statements that use function notation in terms of a context.

Analyze functions using different representations.

F-IF.7. Graph functions expressed symbolically and show key features of the graph, by hand in simple cases and using technology for more complicated cases.★

    c. Graph polynomial functions, identifying zeros when suitable factorizations are available, and showing end behavior.

Interpreting Functions
Set 7: Polynomial Functions

Instruction

Student Activities Overview and Answer Key

Station 1

Working with groups, students use the radical root theorem, the fundamental theorem of algebra, and the remainder theorem to evaluate polynomials and find their factors.

Answers

1. a. 5
   b. no
   c. $\pm\dfrac{7,1}{3,1}$
   d. –5

2. a. 4
   b. $\pm\dfrac{1}{3,1}$
   c. no
   d. no
   e. no

3. a. 3
   b. $\pm\dfrac{1,2,5,10}{1,2}$
   c. no
   d. –290

4. a. 6
   b. ±1
   c. –2
   d. 6

Station 2

Working with groups, students use technology to evaluate polynomials.

Answers

1. –12
2. 59,452
3. –1,354
4. 8
5. 40,905
6. 55,017
7. 36

Interpreting Functions
Set 7: Polynomial Functions

Instruction

Station 3

Working with groups, students work to factor polynomials, including polynomials with radical and complex roots.

Answers

1. a. 7
   b. 6
   c. 2
2. a. 6
   b. 131,503
   c. 3
3. $(x-9)(x+6)(x-2)(x+1)$
4. $(x-2)(x+2)(x+3)(x-1)(x-4)$
5. $\left(x-\sqrt{2}\right)(x-4)\left(x+\dfrac{2}{3}\right)$
6. $-3\sqrt{5}, -\dfrac{1}{4}, 2$
7. $1+2i, 3$
8. $2-2i, 2, -\dfrac{1}{2}$

Interpreting Functions
Set 7: Polynomial Functions

Instruction

Station 4

Working with groups, students use technology to evaluate and graph polynomials. Students also use the rational root theorem to find possible roots.

Answers

1. a. $(x-2)(x+3)(x+2)$

   b.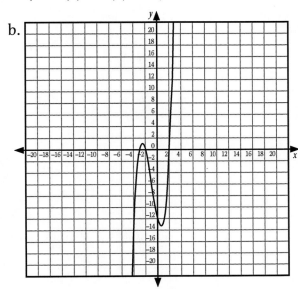

2. a. $(x+5)(x-1)(3x+2)(x-3)$

   b.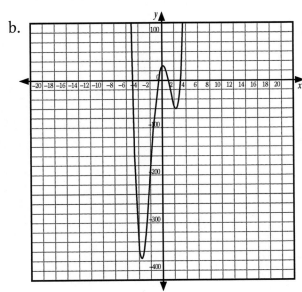

Interpreting Functions
Set 7: Polynomial Functions

Instruction

3. a. $(x + i)(x - i)(x - 4)$

   b.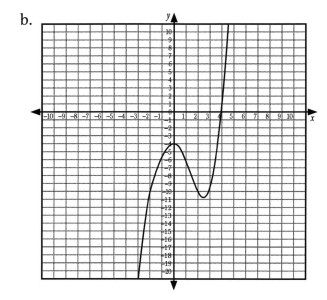

4. a. ±3, 1; there are no rational roots to this function, only irrational and complex.

   b.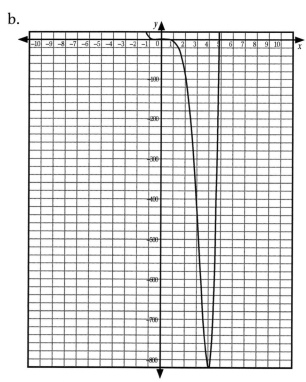

Interpreting Functions
Set 7: Polynomial Functions

Instruction

5. a. $\pm 1, 2, 4, \dfrac{1}{3}, \dfrac{1}{9}, \dfrac{2}{3}, \dfrac{2}{9}, \dfrac{4}{3}, \dfrac{4}{9}$; $x = 1$. The other roots are complex.

   b.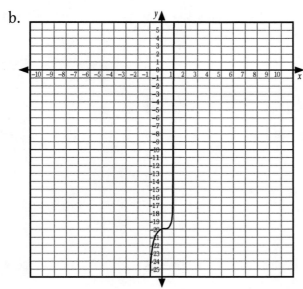

6. a. $\pm 1, 2, 4, \dfrac{1}{3}, \dfrac{1}{9}, \dfrac{2}{3}, \dfrac{2}{9}, \dfrac{4}{3}, \dfrac{4}{9}$; $(3x + i)(3x - i)(x + 2i)(x - 2i)(x + 1)$

   b.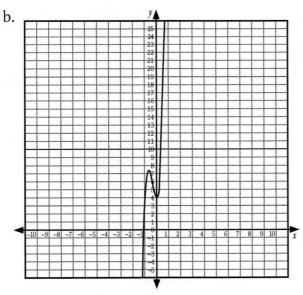

Materials List/Setup

Station 1   none

Station 2   calculator

Station 3   calculator

Station 4   calculator or computer station with graphing software

Interpreting Functions
Set 7: Polynomial Functions

Instruction

Discussion Guide

To support students in reflecting on the activities and to gather some formative information about student learning, use the following prompts to facilitate a class discussion to "debrief" the station activities.

Prompts/Questions
1. What is the degree of a polynomial?
2. What is the fundamental theorem of algebra?
3. What is a complex conjugate? What is a radical conjugate?
4. How are conjugates useful when you are factoring?

Think, Pair, Share

Have students jot down their own responses to questions, then discuss with a partner (who was not in their station group), and then discuss as a whole class.

Suggested Appropriate Responses
1. The degree of a polynomial is the highest power to which x is raised.
2. Any polynomial of degree x^n will have n roots.
3. If you have a complex number in the form $a + bi$, the complex conjugate is $a - bi$. If you have a radical in the form $a + b\sqrt{c}$, the complex conjugate is $a - b\sqrt{c}$.
4. Conjugates allow you to treat $x^2 - n$ as the difference of two squares, even if n is not a perfect square. Additionally, if any polynomial has a complex or radical root, the root's conjugate will also be a root.

Possible Misunderstandings/Mistakes
- Incorrectly calculating the value of functions
- Incorrectly manipulating exponents
- Incorrectly factoring
- Not understanding the rational root theorem
- Not extending the rational root theorem far enough (i.e., not listing all possible factors of the constant and coefficient)

Interpreting Functions
Set 7: Polynomial Functions

Instruction

- Not recognizing redundant factors in the list of possibilities from the rational root theorem
- Incorrectly using graphing technology
- Attempting to find zeros from the polynomial's graph
- Not understanding or misusing polynomial long division and synthetic long division
- Not understanding the remainder theorem
- Making arithmetical errors

Interpreting Functions
Set 7: Polynomial Functions

Station 1

Work with your group to answer the questions about each polynomial function.

1. $f(x) = 3x^5 + x^3 - 2x^2 + 6x + 7$
   a. How many roots does this function have?
   b. Is 3 a zero?
   c. What are all possible rational zeros?
   d. What is $f(-1)$?

2. $f(x) = 3x^4 + 2x + 1$
   a. How many roots does this function have?
   b. What are all possible rational zeros?
   c. Is $\frac{1}{3}$ a zero?
   d. Is $-\frac{1}{3}$ a zero?
   e. Is -1 a zero?

3. $g(x) = -2x^3 - 3x^2 + 5x + 10$
   a. How many roots does this function have?
   b. What are all possible rational zeros?
   c. Is 1 a zero?
   d. What is $g(5)$?

4. $g(x) = x^6 + 4x^5 + 2x^4 - 1$
   a. How many roots does this function have?
   b. What are all possible rational zeros?
   c. What is $g(-1)$?
   d. What is $g(1)$?

NAME:

Interpreting Functions
Set 7: Polynomial Functions

Station 2

Work with your group to answer the questions about each polynomial function. Use a calculator to compute values.

1. $g(x) = -2x^3 + 2x^2 - 7x + 10$

    $g(2) =$

2. $f(x) = 4x^7 - 6x^5 + x^4 - 3x^3 - 4$

    $f(4) =$

3. $f(x) = -2x^8 - 7x^7 + x^6 - 2x^5 + x^4 + 5x^3 - x$

    $f(2) =$

4. $g(x) = x^4 - 5x + 2$

    $g(2) =$

5. $f(x) = 3x^6 - 2x^5 + x^4 - 2x^3 - 4x^2 + 5$

    $f(5) =$

6. $g(x) = 18x^5 - 10x^3 + 2x^2 - 6x - 3$

    $g(5) =$

7. $h(x) = 4x^4 - 3x^3 + 2x^2 - x - 10$

    $h(2) =$

Interpreting Functions
Set 7: Polynomial Functions

Station 3

Work with your group to answer the questions about each polynomial function. Show all your work.

1. $g(x) = 4x^7 - 2x + 4$

   a. How many roots does the function have?
   b. What is $g(1)$?
   c. What is $g(-1)$?

2. $f(x) = 2x^6 + 5x^5 - 3x^3 - x^2 - x + 1$

   a. How many roots does the function have?
   b. What is $f(6)$?
   c. What is $f(1)$?

Factor the following polynomials. Use any method you like.

3. $x^4 - 4x^3 - 53x^2 + 60x + 108 = 0$

4. $x^5 - 2x^4 - 15x^3 + 20x^2 + 44x - 48 = 0$

5. If $\left(x + \sqrt{2}\right)$ is a factor of $3x^4 - 10x^3 - 14x^2 + 20x + 16 = 0$, what are the other factors?

6. If $3\sqrt{5}$ is a root of $4x^4 - 7x^3 - 182x^2 + 315x + 90 = 0$, what are the other roots?

7. If $(1 - 2i)$ is a root of $x^3 - 5x^2 + 11x - 15 = 0$, what are the other roots?

8. If $(2 + 2i)$ is a root of $2x^4 - 11x^3 + 26x^2 - 16x - 16 = 0$, what are the other roots?

Interpreting Functions
Set 7: Polynomial Functions

Station 4

Work with your group to answer the questions about each polynomial function. Then use a graphing calculator or graphing software to find the graph of the function. Sketch the graphs.

1. $x^3 + 3x^2 - 4x - 12 = 0$

   a. What are the factors?

   b.
   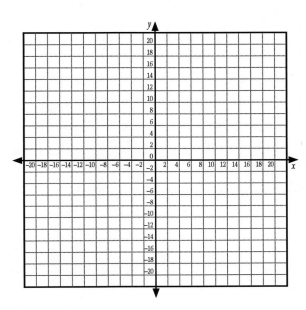

2. $3x^4 + 5x^3 - 49x^2 + 11x + 30 = 0$

   a. What are the factors?

   b.
   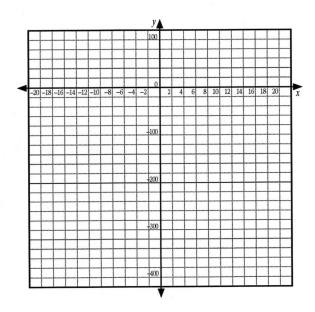

continued

Interpreting Functions
Set 7: Polynomial Functions

3. $x^3 - 4x^2 - 2x + 8 = 0$

   a. If $-i$ is one root, what are all the factors?

   b.
   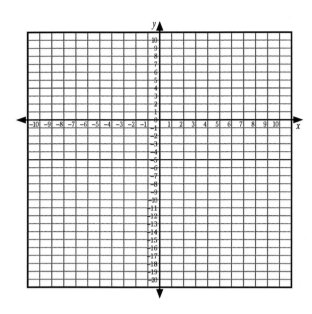

4. $x^6 - 5x^5 + x^4 - 3x^2 + 3 = 0$

   a. State all possible rational roots and find all rational roots.

   b.
   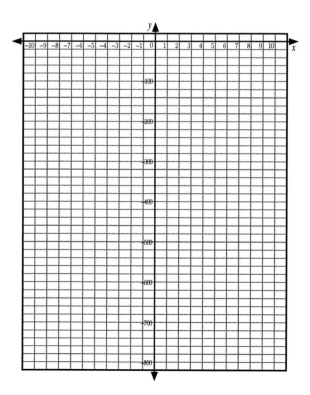

continued

Interpreting Functions
Set 7: Polynomial Functions

5. $10x^5 + 19x^3 - 10x^2 + x - 20 = 0$

   a. State all possible rational roots and find all rational roots.

   b.
   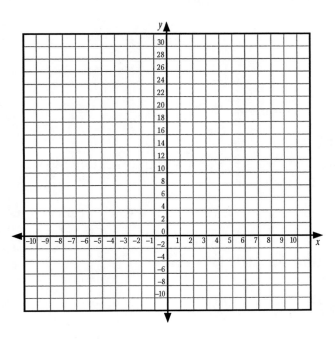

6. $9x^5 + 9x^4 + 37x^3 + 37x^2 + 4x + 4 = 0$

   a. State all possible rational roots and find ALL roots. (*Hint*: use the quadratic formula after finding a rational root.)

   b.
   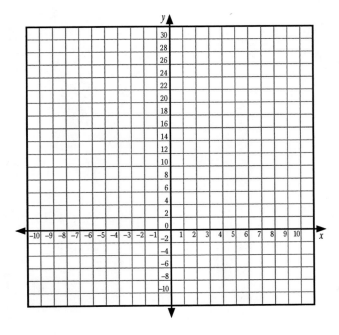

Building Functions

Set 1: Inverse Functions

Instruction

Goal: To guide students to an understanding of inverse functions, including finding and graphing a function's inverse

Common Core Standards

Functions: Building Functions

Build new functions from existing functions.

F-BF.4. Find inverse functions.

Solve an equation of the form $f(x) = c$ for a simple function f that has an inverse and write an expression for the inverse.

(+) Verify by composition that one function is the inverse of another.

Student Activities Overview and Answer Key

Station 1

Working with groups, students find the inverses of various functions.

Answers

1. c
2. c
3. c
4. b
5. $f(f^{-1}(x)) = 3 - \left(\dfrac{-2x+6}{2}\right)$
   $= 3 + x - 3$
   $= x$
6. $f(f^{-1}(x)) = \left((x+2)^{\frac{1}{3}}\right)^3 - 2$
   $= x + 2 - 2$
   $= x$

Building Functions
Set 1: Inverse Functions

Instruction

7. $f(f^{-1}(x)) = \left(\dfrac{1}{\dfrac{2}{(2x-6)}}\right) + 3$

$= \dfrac{2x-6}{2} + 3$

$= x - 3 + 3$

$= x$

Station 2

Working with partners, students find the inverses of linear functions and graph each function with its inverse.

Answers

1. $f^{-1}(x) = \dfrac{2}{5}(x-2)$

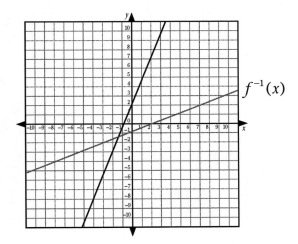

2. $f^{-1}(x) = \dfrac{1}{3}(x+1)$

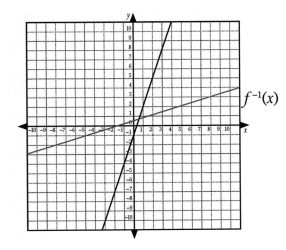

Building Functions
Set 1: Inverse Functions

Instruction

3. $f^{-1}(x) = \dfrac{1}{2}(x-5)$

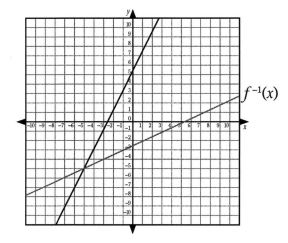

4. $f^{-1}(x) = \dfrac{1}{3}x - \dfrac{1}{6}$

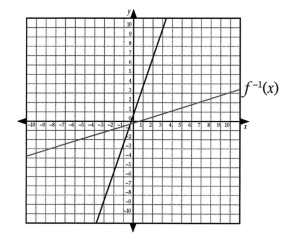

5. $f^{-1}(x) = \dfrac{1}{4}x + \dfrac{1}{2}$

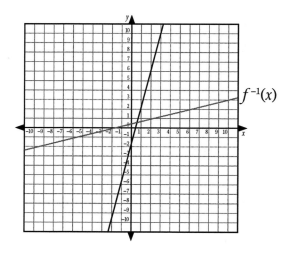

Building Functions
Set 1: Inverse Functions

Instruction

Station 3

Working with groups, students find the inverses of quadratic and power functions and graph each function with its inverse. They also assess whether each function is one-to-one.

Answers

1. a. $f^{-1}(x) = \pm\sqrt{x} - 3$

   b.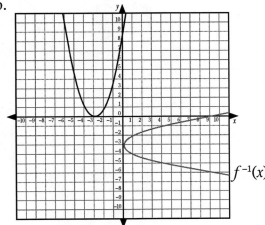

   c. no

2. a. $f^{-1}(x) = (x+4)^{\frac{1}{3}}$

   b.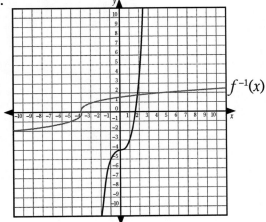

   c. yes

Building Functions
Set 1: Inverse Functions

Instruction

3. a. $f^{-1}(x) = (2x)^{\frac{1}{5}}$
   b.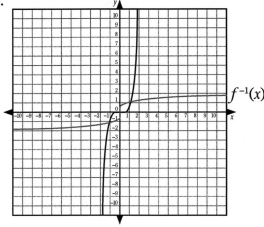
   c. yes

4. a. $f^{-1}(x) = \pm \dfrac{\sqrt{2x}}{2}$
   b.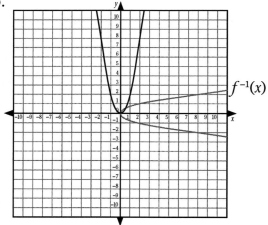
   c. no

Building Functions
Set 1: Inverse Functions

Instruction

Station 4

Working with groups, students find the inverses of functions in the form $f(x) = \dfrac{a}{x}$ and graph each function with its inverse. They also assess whether each function is one-to-one.

Answers

1. a. $f^{-1}(x) = \dfrac{2}{x}$

    b.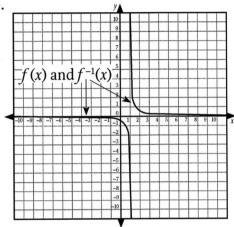

    c. yes

2. a. $f^{-1}(x) = \dfrac{3}{x} + 2$

    b.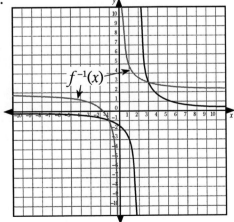

    c. yes

Building Functions
Set 1: Inverse Functions

Instruction

3. a. $f^{-1}(x) = \dfrac{1}{4(x-1)}$

   b.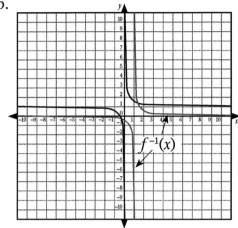

   c. yes

Materials List/Setup

Station 1   none
Station 2   graphing calculator
Station 3   none
Station 4   none

Building Functions
Set 1: Inverse Functions

Instruction

Discussion Guide

To support students in reflecting on the activities and to gather some formative information about student learning, use the following prompts to facilitate a class discussion to "debrief" the station activities.

Prompts/Questions

1. What is a function?
2. What is a reflection?
3. Why might it be important to know the inverse of a function?
4. What is a one-to-one function?

Think, Pair, Share

Have students jot down their own responses to questions. Pair students from different station groups and have them discuss their answers. Finally, discuss answers as a whole class.

Suggested Appropriate Responses

1. A function is a description of a mathematical operation in which each value in the domain is mapped to exactly one value in the range. The vertical line test can be used to determine if a graph represents a function.
2. A reflection is the graph of the function across an axis of symmetry. Each point on the graph is the same distance from the axis of symmetry as the corresponding point on the reflection.
3. to perform additional algebraic operations; to "undo" what a function has done
4. a function in which each output value is mapped to one and only one input value

Building Functions
Set 1: Inverse Functions

Instruction

Possible Misunderstandings/Mistakes

- Not understanding how a function's inverse works
- Not understanding the steps of finding a function's inverse
- Incorrectly manipulating numbers, variables, or exponents
- Not understanding how to use composition to verify the inverse of a function
- Confusing the inverse with the conjugate
- Not understanding the relationship between whole-number exponents and their fraction inverses
- Not understanding the vertical-line test
- Not understanding the relationship between inverses and if they represent a one-to-one function

Building Functions
Set 1: Inverse Functions

Station 1

Work with your partner to answer each question. Show your work.

For problems 1–4, find the inverse of each given function. Circle the letter of the correct answer.

1. $f(x) = 3x + 2$

   a. $\dfrac{1}{3x-2}$

   b. $3x + 2$

   c. $\dfrac{x-2}{3}$

2. $f(x) = x^2 + 3$

   a. $\dfrac{1}{x^2+3}$

   b. $3x + 2$

   c. $\pm\sqrt{x-3}$

3. $f(x) = \dfrac{x^4}{2}$

   a. $2x^{\frac{1}{4}}$

   b. $\dfrac{x^2}{4}$

   c. $\pm(2x)^{\frac{1}{4}}$

continued

Building Functions
Set 1: Inverse Functions

4. $f(x) = \dfrac{3}{x}$

   a. $\dfrac{x}{3}$

   b. $\dfrac{3}{x}$

   c. x^3

5. If $f(x) = 3 - \dfrac{x}{2}$, prove that $f^{-1}(x) = -2x + 6$.

6. If $f(x) = x^3 - 2$, prove that $f^{-1}(x) = (x+2)^{\frac{1}{3}}$.

7. If $f(x) = \dfrac{1}{2x} + 3$, prove that $f^{-1}(x) = \dfrac{1}{2x - 6}$.

Building Functions
Set 1: Inverse Functions

Station 2

Work with your partner to find and graph the inverse of the given function. Show all your work. Use the calculator if you need help.

1. $f(x) = \dfrac{5}{2}x + 2$

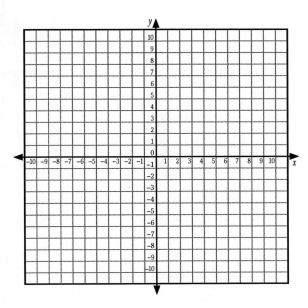

2. $f(x) = 3x - 1$

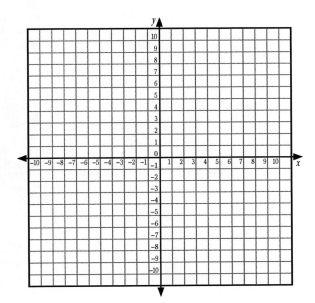

continued

Building Functions
Set 1: Inverse Functions

3. $f(x) = 2x + 5$

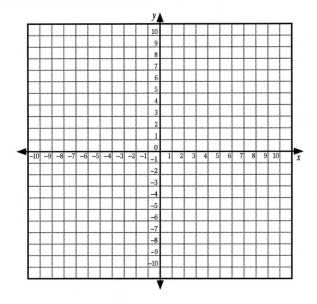

4. $f(x) = 3x + \dfrac{1}{2}$

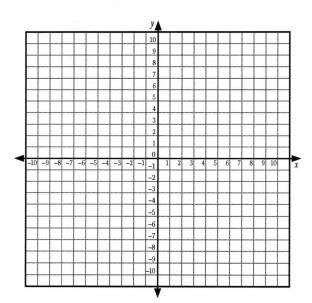

continued

Building Functions
Set 1: Inverse Functions

5. $f(x) = 4x - 2$

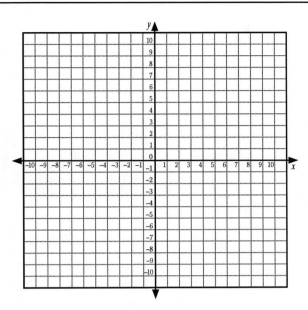

Building Functions
Set 1: Inverse Functions

Station 3

Work with your group. For each function, a) provide the inverse; b) graph the function and its inverse; and c) state whether the function is one-to-one. Show all your work.

1. $f(x) = (x+3)^2$

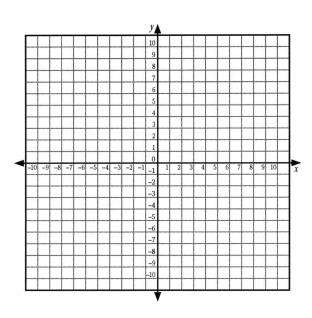

2. $f(x) = x^3 - 4$

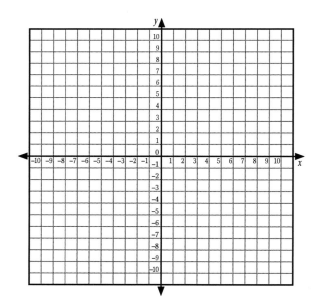

continued

Building Functions
Set 1: Inverse Functions

3. $f(x) = \dfrac{x^5}{2}$

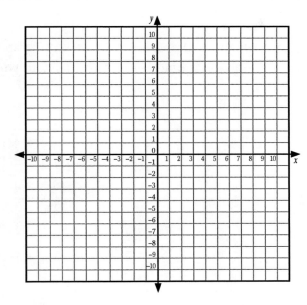

4. $f(x) = 2x^2$

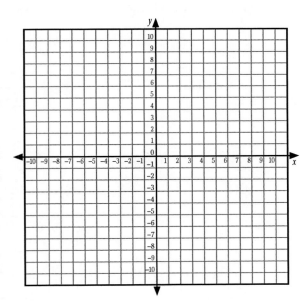

Building Functions
Set 1: Inverse Functions

Station 4

Work with your group. For each function, a) provide the inverse; b) graph the function and its inverse; and c) state whether the function is one-to-one. Show all your work.

1. $f(x) = \dfrac{2}{x}$

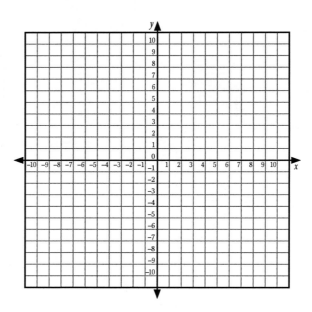

2. $f(x) = \dfrac{3}{x-2}$

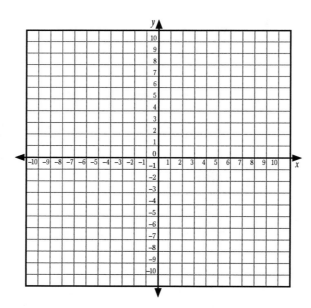

continued

Building Functions
Set 1: Inverse Functions

3. $f(x) = \dfrac{1}{4x} + 1$

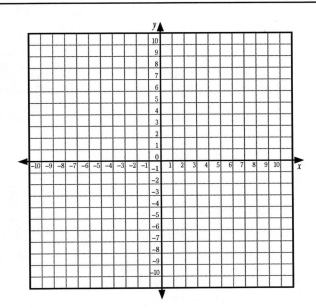

Building Functions

Set 2: Logarithmic Functions As Inverses of Exponential Functions

Instruction

Goal: To guide students to an understanding of the inverse relationship between logarithms and exponential functions, and of how to solve exponential equations and application problems using common and natural logarithms

Common Core Standards

Functions: Interpreting Functions

Analyze functions using different representations.

F-IF.7. Graph functions expressed symbolically and show key features of the graph, by hand in simple cases and using technology for more complicated cases.★

e. Graph exponential and logarithmic functions, showing intercepts and end behavior, and trigonometric functions, showing period, midline, and amplitude.

Functions: Building Functions

Build new functions from existing functions.

F-BF.5. (+) Understand the inverse relationship between exponents and logarithms and use this relationship to solve problems involving logarithms and exponents.

Student Activities Overview and Answer Key

Station 1

Working with partners, students graph exponential functions. Students find the inverse of each function and graph it. A suggested extension activity is to have students identify the domain and range of the function and its inverse and then make comparisons.

Building Functions
Set 2: Logarithmic Functions As Inverses of Exponential Functions

Instruction

Answers

1. $y = \log_3(x) - 1$

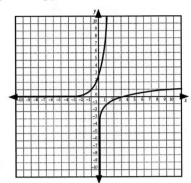

2. $y = \log_2(x)$

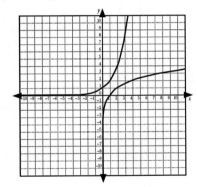

3. $y = \log_{10}(x + 4)$

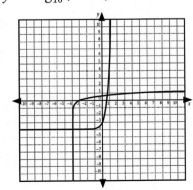

4. $y = \log_{\frac{1}{2}}(x)$

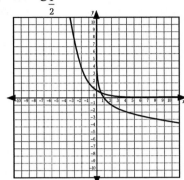

5. $y = \log_4 x + 2$

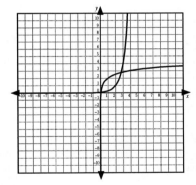

6. $y = 2\log_5 x$

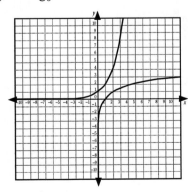

Building Functions
Set 2: Logarithmic Functions As Inverses of Exponential Functions

Instruction

Station 2

Working with groups, students expand and simplify logarithmic solutions to exponential equations. Students either leave the answers in logarithm form or use common logarithms to solve the equations. An alternative is to create cards with the expression and solutions and have students match the cards.

Answers

1. 4
2. –4
3. 5
4. $\log_4 5 + 2\log_4 x - \dfrac{1}{2}$
5. $3\log_2(x+3) - \log_2(x) - 2\log_2(x-2)$
6. $b = 10$
7. $x = 2.27375$
8. $x = 2.321928$

Station 3

Working with groups, students use natural logarithms and calculators to solve application problems (compounding interest) with exponential equations.

Answers

1. 12.79 years
2. 25.56 years
3. 46.38 years
4. 7.81 years
5. 12.54 years

Building Functions
Set 2: Logarithmic Functions As Inverses of Exponential Functions

Instruction

Station 4

Working with groups, students solve application problems (growth and decay) with exponential equations of base *e*.

Answers

1. 5.1344 years
2. 778.364 years
3. $r = -0.003465$
4. 3.984 hours
5. $r = 0.26483$
6. 53.908 hours

Materials List/Setup

Station 1	colored pens or pencils
Station 2	calculator
Station 3	calculator
Station 4	calculator

Building Functions
Set 2: Logarithmic Functions As Inverses of Exponential Functions

Instruction

Discussion Guide

To support students in reflecting on the activities and to gather some formative information about student learning, use the following prompts to facilitate a class discussion to "debrief" the station activities.

Prompts/Questions

1. What is a logarithm?
2. How does the base of a logarithm relate to an exponential function?
3. What is e?
4. What is the difference between a common logarithm and a natural logarithm?

Think, Pair, Share

Have students jot down their own responses to questions, then discuss with a partner (who was not in the same station group), and then discuss as a whole class.

Suggested Appropriate Responses

1. A logarithm is the inverse of an exponential function.
2. The base of a logarithm is the base of the corresponding exponential function. If $\log_b(y) = x$, then $y = b^x$.
3. e is an irrational number that applies to any situation in which there is continuous growth or decay.
4. A common logarithm has base 10. A natural logarithm has base e.

Possible Misunderstandings/Mistakes

- Incorrectly manipulating numbers, variables, or exponents
- Not understanding the laws of exponents
- Not understanding the difference between growth and decay
- Incorrectly calculating squares, cubes, etc., of integers between 1 and 10
- Incorrectly using the exponent function of a calculator
- Incorrectly using the log or ln function of a calculator
- Incorrectly applying the formula of compound interest

Building Functions
Set 2: Logarithmic Functions As Inverses of Exponential Functions

Instruction

- Not understanding the difference between total time t and intervals m
- Incorrectly applying the growth/decay formula
- Not understanding the relationship between an exponential function and its graph
- Not understanding the relationship between an exponential function and its logarithm
- Not understanding the difference between common and natural logarithms
- Using the numeric estimate of e and calculating a natural logarithm arithmetically, rather than simplifying it
- Rounding off numbers too early in calculations

Building Functions

Set 2: Logarithmic Functions As Inverses of Exponential Functions

Station 1

Work with your partner to graph each function, then to find and graph its inverse. Use graph paper. Write the equation of the inverse next to the original function.

1. $y = 3^{x+1}$

2. $y = 2^x$

3. $y = 10^x - 4$

4. $y = \left(\dfrac{1}{2}\right)^x$

5. $\log_4(y) = x - 2$

6. $\log_5(y) = \dfrac{x}{2}$

Building Functions
Set 2: Logarithmic Functions As Inverses of Exponential Functions

Station 2

Work with your group to solve each problem.

1. $\log_3(81) =$

2. $\log_5\left(\dfrac{1}{625}\right) =$

3. $\log_2(32) =$

Expand each expression.

4. $\log_4\left(\dfrac{5x^2}{2}\right) =$

5. $\log_2\left(\dfrac{(x+3)^3}{x(x-2)^2}\right) =$

continued

Building Functions
Set 2: Logarithmic Functions As Inverses of Exponential Functions

Solve for b.

6. $\log_b(100) = 2$
$\log_b(1,000,000) = 6$
$b =$

Use common logarithms to solve the next two equations.

7. $2^{x^2+1} = 72$

8. An element decays according to the formula $Q = Q_0\left(1 - \dfrac{1}{2}\right)^x$, where x = time. If the initial amount (Q_0) is 50 mg, how long will it take to decay to 10 mg?

Building Functions
Set 2: Logarithmic Functions As Inverses of Exponential Functions

Station 3

Work with your group to answer each question, using natural logarithms. Use the formula $A = P\left(1 + \dfrac{r}{m}\right)^{tm}$ to calculate compound interest. Use the calculator to find the final values. Round decimals to two places.

1. In a bank account, interest compounds at 4% monthly. After how many years would a deposit of $600 grow to $1000?

2. In a bank account, interest compounds at 2% monthly. After how many years would a deposit of $300 grow to $500?

3. In a bank account, interest compounds at 3% quarterly. After how many years would a deposit of $2500 grow to $10,000?

4. In a bank account, interest compounds at 5.2% monthly. After how many years would a deposit of $1000 grow to $1500?

5. In a bank account, interest compounds at 2.3% quarterly. After how many years would a deposit of $7500 grow to $10,000?

Building Functions
Set 2: Logarithmic Functions As Inverses of Exponential Functions

Station 4

Work with your group to set up and then calculate each equation. To calculate growth or decay, use the formula $Q = Q_0(e)^{rt}$, where Q_0 is the initial amount, r is the rate of growth or decay, t is the amount of time that passes, and Q is the amount after t.

1. A radioactive element decays at a rate of $r = -0.135$. What is its half-life? (An element's half-life is the amount of time it takes for an amount of the element to decay to half of the original amount.)

2. An element decays at a rate of $r = -0.0025$. How long, in years, will it take a quantity of 70 mg to decay to 10 mg?

3. If Element A has a half-life of 200 years, what is its rate of decay?

4. A colony of bacteria starts with a population of 250. It grows at a rate of $r = 0.752$. How many hours will it take for the population to reach 5000?

5. A colony of bacteria starts with a population of 500. After 12 hours, it has a population of 12,000. What is its rate of growth?

6. A colony of bacteria starts with a population of 20. It grows at a rate of $r = 0.124$. How many hours will it take for the population to reach 16,000?

Linear, Quadratic, and Exponential Models

Set 1: Arithmetic Sequences and Series

Instruction

Goal: To guide students to an understanding of arithmetic sequences and series, including partial sums as examples of quadratic functions

Common Core Standards

Functions: Interpreting Functions

Understand the concept of a function and use function notation.

F-IF.3. Recognize that sequences are functions, sometimes defined recursively, whose domain is a subset of the integers.

Functions: Building Functions

Build a function that models a relationship between two quantities.

F-BF.2. Write arithmetic and geometric sequences both recursively and with an explicit formula, use them to model situations, and translate between the two forms.★

Functions: Linear, Quadratic, and Exponential Models★

Construct and compare linear, quadratic, and exponential models and solve problems.

F-LE.2. Construct linear and exponential functions, including arithmetic and geometric sequences, given a graph, a description of a relationship, or two input-output pairs (include reading these from a table).

Student Activities Overview and Answer Key

Station 1

Working with partners, students find the difference d or the nth term a_n of various arithmetic sequences.

Answers

1. $d = 4$
2. $d = -3$
3. $a_9 = 16$
4. $a_{10} = 6$
5. $d = 3$
6. $a_{10} = 3x + 36$
7. $a_8 = -5x$
8. Answers will vary. Students should find a_7 of their partners' sequences.

Linear, Quadratic, and Exponential Models
Set 1: Arithmetic Sequences and Series

Instruction

Station 2

Working with partners, students find the partial sums of arithmetic sequences.

Answers

1. $a_n = 2t(1 + n)$
2. $a_{10} = 22t$
3. $S_{10} = 130t$
4. $a_n = 10n - 5$
5. $a_{15} = 145$
6. $S_{15} = 1125$
7. $S_9 = 81x$
8. No. There is not a constant difference between the terms.
9. $a_{35} = 69$
10. $S_{35} = 1225$

Station 3

Students work in groups to find the sum, the nth term, the number of terms, and the first term of various series. Students connect sequences in list form with sequences in sigma form.

Answers

1. $x - 2$ terms
2. $13 - x$ terms
3. $x = 15$
4. $x = 12$
5. $x = 20$
6. $x = 2$
7. $x = 3$
8. x must always be a positive integer because it refers to a concrete, countable quantity, the number of terms in the series.

Linear, Quadratic, and Exponential Models
Set 1: Arithmetic Sequences and Series

Instruction

Station 4

Students work in groups to answer questions about a sequence and its related series, as well as the graphs of each function. Students are guided to recognize that the sequence can be expressed as a linear function, whereas the series can be expressed as a quadratic function.

Answers

1. $n = 7$

2.
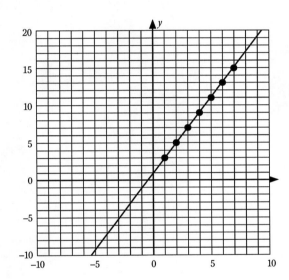

3. $y = 2x + 1$

4.
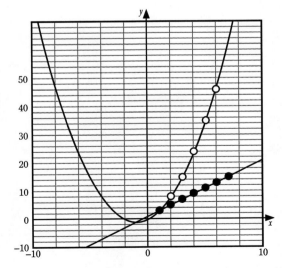

Note: Graph of parabola added to show shape only

Linear, Quadratic, and Exponential Models
Set 1: Arithmetic Sequences and Series

Instruction

5. No. It's quadratic.

6. $\sum_{x=1}^{7} x^2 + 2x$

Materials List/Setup

Station 1   none
Station 2   calculator
Station 3   calculator
Station 4   ruler; calculator

Linear, Quadratic, and Exponential Models
Set 1: Arithmetic Sequences and Series

Instruction

Discussion Guide

To support students in reflecting on the activities and to gather some formative information about student learning, use the following prompts to facilitate a class discussion to "debrief" the station activities.

Prompts/Questions

1. What is an arithmetic sequence?
2. What is a series?
3. What are some ways of showing a series?
4. Is a sequence finite or infinite? Is a series finite or infinite?

Think, Pair, Share

Have students jot down their own responses to questions, then discuss with a partner (who was not in their station group), and then discuss as a whole class.

Suggested Appropriate Responses

1. An arithmetic sequence is an ordered group of numbers separated by a common difference.
2. A series is the partial sum of a sequence.
3. One way is with a $\sum$ formula, with the expression S_n, in which S stands for the sum and n stands for the number of terms included.
4. A sequence can be bounded (finite) or infinite. An arithmetic series must be finite because you cannot find an infinite sum.

Possible Misunderstandings/Mistakes

- Not understanding sigma notation of series
- Not understanding the difference between sequences and series
- Not understanding the difference between arithmetical and geometric sequences
- Incorrectly finding the common difference d
- Making simple arithmetical errors
- Assuming that S_n includes n terms, even if the series does not begin at a_1
- Not understanding the difference between linear and quadratic expressions

Linear, Quadratic, and Exponential Models
Set 1: Arithmetic Sequences and Series

Station 1

Work with your partner to answer each question. Show your work.

1. Find the difference d of the sequence 1, 5, 9, 13, 17, ...

2. Find the difference d of the sequence 2, −1, −4, −7, −10, ...

3. Find the ninth term a_9 of the sequence 0, 2, 4, 6, 8, ...

4. Find the tenth term a_{10} of the sequence $\frac{3}{2}, 2, \frac{5}{2}, 3, \frac{7}{2}, ...$

5. Find the common difference of the sequence $a_n = 3n + 2$.

6. Find a_{10} for the sequence $3x, 3x + 4, 3x + 8, 3x + 12, ...$

7. Find a_8 for $2x, x, 0, -x, -2x, ...$

8. Create a sequence for your partner. Have your partner find a_7.

Linear, Quadratic, and Exponential Models
Set 1: Arithmetic Sequences and Series

Station 2

Work with your partner to answer each question. Show all your work. Use the calculator if you need help.

1. $a_1 = 4t$
   $d = 2t$
   What is a_n?

2. What is a_{10}?

3. What is the sum of the first 10 terms in the sequence?

4. $a_1 = 5$
   $d = 10$
   What is a_n?

5. What is a_{15}?

6. What is S_{15}?

7. Look at the sequence $x, 3x, 5x, 7x, \ldots$
   What is S_9?

8. Look at the sequence 0, 1, 4, 9, 16, 25, …
   Is it arithmetical? Explain.

9. Think of the sequence of positive odd integers. What is the 35th term of that sequence?

10. What is S_{35} of the sequence of positive odd integers?

Linear, Quadratic, and Exponential Models
Set 1: Arithmetic Sequences and Series

Station 3

Work with your group to solve the problems. Use a calculator if you need to. Show all your work.

1. In terms of x, state the number of terms in the series.

$$\sum_{a=3}^{x} 2a + 1$$

2. In terms of x, state the number of terms in the series.

$$\sum_{a=x}^{12} 2a + 1$$

Solve problems 3–7 for x.

3. $$\sum_{a=1}^{x} 4a + 2 = 510$$

4. $$\sum_{a=4}^{x} \frac{1}{2}a - 1 = 27$$

5. $$\sum_{a=5}^{x} 3a + 3 = 648$$

6. $$\sum_{a=x}^{10} 2a - 1 = 99$$

7. $$\sum_{a=x}^{15} \frac{x}{3} + 2 = 65$$

8. Explain in words why in any series $\sum_{a=1}^{x}$, x must always be a positive integer.

Linear, Quadratic, and Exponential Models
Set 1: Arithmetic Sequences and Series

Station 4

Work with your group to solve each problem. Show all your work.

1. Solve for n.

$$\sum_{a=1}^{n} 2a + 1 = 63$$

2. Look at the sequence expressed in problem 1. Graph it as a line on the system below.

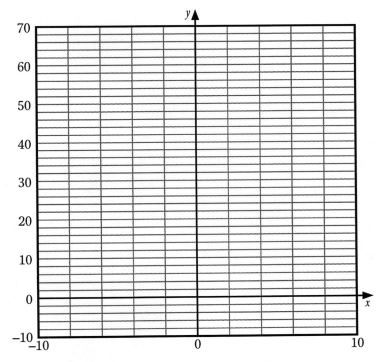

3. Write the equation of the sequence as a line in the form $y = mx + b$.

4. Now, graph the series from problem 1 on the same graph.

5. Is this graph linear? Explain.

6. Write the equation of the series for all x values between $x = 1$ and $x = 7$ as a quadratic function.

Linear, Quadratic, and Exponential Models

Set 2: Geometric Sequences

Instruction

Goal: To guide students to an understanding of the inverse relationship between logarithms and exponential functions, and of how to solve exponential equations and application problems using common and natural logarithms

Common Core Standards

Functions: Building Functions

Build a function that models a relationship between two quantities.

F-BF.2. Write arithmetic and geometric sequences both recursively and with an explicit formula, use them to model situations, and translate between the two forms.

Functions

Linear, Quadratic, and Exponential Models★

Construct and compare linear, quadratic, and exponential models and solve problems.

F-LE.2. Construct linear and exponential functions, including arithmetic and geometric sequences, given a graph, a description of a relationship, or two input-output pairs (include reading these from a table).

Student Activities Overview and Answer Key

Station 1

Working with partners, students find the common ratios, nth terms, and formulas of geometric sequences.

Answers

1. $r = 2$

    $a_n = 2^{n-1}$

2. 524,288

3. $r = 3$

    $a_n = 2(3^{n-1})$

4. 39,366

5. $r = 6$

    $a_n = 5(6^{n-1})$

6. 233,280

Linear, Quadratic, and Exponential Models
Set 2: Geometric Sequences

Instruction

7. $a_1 = 4$
   $r = 2$

8. $a_1 = 10$
   $r = 6$

Station 2

Working with groups, students graph geometric sequences as exponential functions.

Answers

1.
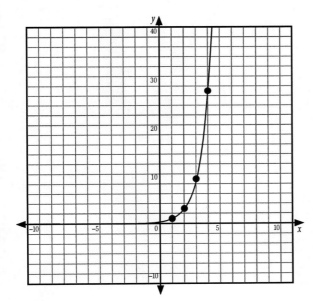

2.
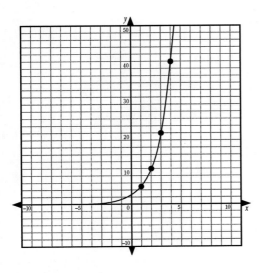

Linear, Quadratic, and Exponential Models
Set 2: Geometric Sequences

Instruction

3.

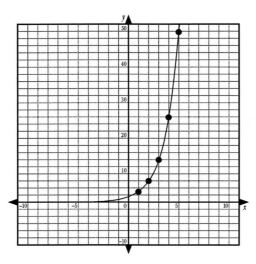

4.

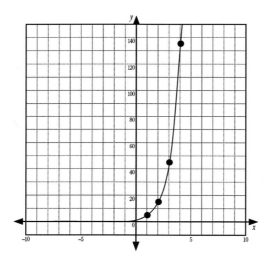

Linear, Quadratic, and Exponential Models
Set 2: Geometric Sequences

Instruction

5.

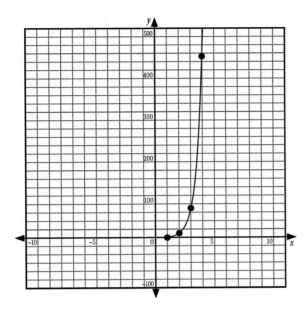

Station 3

Working with groups, students use exponential equations to answer questions about geometric sequences.

Answers

1. $n = 5$
2. $n = 10$
3. $n = 9$
4. $n = 8$
5. $n = 11$
6. To find n, set up an exponential equation in the form $a_n = a_1\left(r^{n-1}\right)$. Using natural logarithms to get n out of the exponent position greatly simplifies the calculation.

Linear, Quadratic, and Exponential Models
Set 2: Geometric Sequences

Instruction

Station 4

Working with partners, students use natural logarithms to answer questions about geometric sequences. Students should see the utility of logarithms and exponential notation.

Answers

1. $n = 9$
2. $n = 11$
3. $n = 9$
4. $n = 10$
5. $n = 9$
6. $n = 7$

Materials List/Setup

Station 1 calculator
Station 2 calculator; colored pens or pencils
Station 3 calculator
Station 4 calculator

Linear, Quadratic, and Exponential Models
Set 2: Geometric Sequences

Instruction

Discussion Guide

To support students in reflecting on the activities and to gather some formative information about student learning, use the following prompts to facilitate a class discussion to "debrief" the station activities.

Prompts/Questions

1. What is a geometric sequence?
2. How is a geometric sequence different from an arithmetic sequence?
3. How could a geometric sequence be related to exponential functions?
4. How could you use logarithms to answer questions about a geometric sequence?

Think, Pair, Share

Have students jot down their own responses to questions, then discuss with a partner (who was not in their station group), and then discuss as a whole class.

Suggested Appropriate Responses

1. A geometric sequence is an ordered set of numbers that increase or decrease at a common ratio r.
2. An arithmetic sequence is an ordered set of numbers that increase or decrease at a common difference d. The terms in an arithmetic sequence are defined by addition or subtraction; the terms in a geometric sequence are determined by an exponential function.
3. To find the terms of a geometric sequence, we use an exponential calculation.
4. Set up the geometric sequence as an exponential equation, and use logarithms to solve the equation.

Linear, Quadratic, and Exponential Models
Set 2: Geometric Sequences

Instruction

Possible Misunderstandings/Mistakes
- Incorrectly manipulating numbers, variables, or exponents
- Not understanding the laws of exponents
- Incorrectly calculating squares, cubes, etc., of integers between 1 and 10
- Incorrectly using the exponent function of a calculator
- Incorrectly using the ln function of a calculator
- Not understanding the relationship between a geometric sequence and an exponential function
- Not understanding the relationship between an exponential function and its graph
- Not understanding the relationship between an exponential function and its logarithm
- Rounding off numbers too early in calculations
- Confusing geometric and arithmetic sequences
- Confusing sequences with series
- Incorrectly applying the geometric sequence formula

Linear, Quadratic, and Exponential Models
Set 2: Geometric Sequences

Station 1

Work with your partner to answer each question. Use the calculator as you need it.

1. Find r and state the formula of the geometric sequence 1, 2, 4, 8, 16, . . .

2. What is the 20th term of the sequence in problem 1?

3. Find r and state the formula of the geometric sequence 2, 6, 18, 54, 162, . . .

4. What is the 10th term of the sequence in problem 3?

5. Find r and state the formula of the geometric sequence 5, 30, 180, 1080, . . .

6. What is the 7th term of the sequence in problem 5?

7. Find a_1 and r of the geometric sequence with $a_4 = 32$ and $a_9 = 1024$.

8. Find a_1 and r of the geometric sequence with $a_3 = 360$ and $a_9 = 16{,}796{,}160$.

Linear, Quadratic, and Exponential Models
Set 2: Geometric Sequences

Station 2

Working with your group, graph each geometric sequence as an exponential function.

1. 1, 3, 9, 27, . . .

    $a_1 = 1$

    $r = 3$

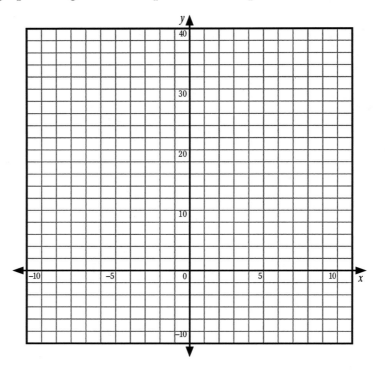

2. 5, 10, 20, 40, . . .

    $a_1 = 5$

    $r = 2$

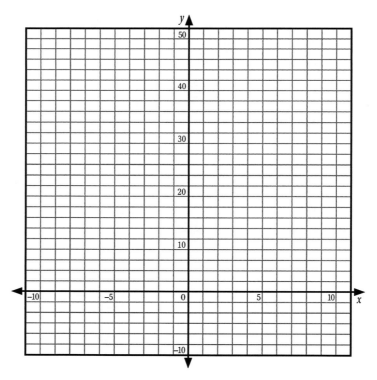

continued

Linear, Quadratic, and Exponential Models
Set 2: Geometric Sequences

3. 3, 6, 12, 24, 48, . . .

   $a_1 = 3$

   $r = 2$

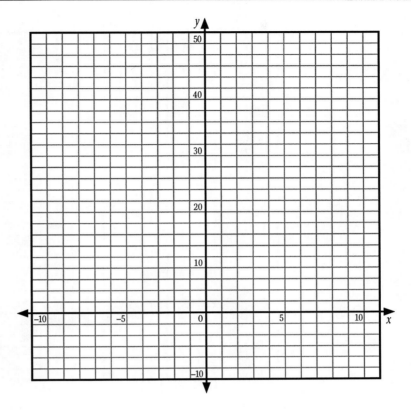

4. 5, 15, 45, 135, . . .

   $a_1 = 5$

   $r = 3$

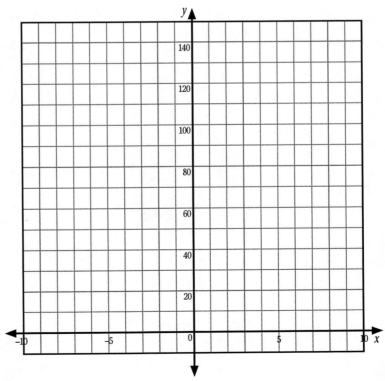

continued

Linear, Quadratic, and Exponential Models
Set 2: Geometric Sequences

5. 2, 12, 72, 432, . . .

$a_1 = 2$

$r = 6$

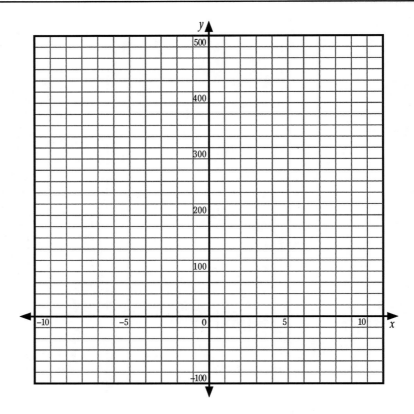

Linear, Quadratic, and Exponential Models
Set 2: Geometric Sequences

Station 3

Work with your group to answer each question. Use a calculator as needed. Show all your work.

1. If some term a_n of a geometric sequence is 48, and $a_1 = 3$ and $r = 2$, what is n?

2. If some term a_n of a geometric sequence is 1,310,720, and $a_1 = 5$ and $r = 4$, what is n?

3. If some term a_n of a geometric sequence is 13,122, and $a_1 = 2$ and $r = 3$, what is n?

4. If some term a_n of a geometric sequence is 384, and $a_1 = 3$ and $r = 2$, what is n?

5. If some term a_n of a geometric sequence is 181,398,528, and $a_1 = 3$ and $r = 6$, what is n?

6. How could you use natural logarithms to find n?

Linear, Quadratic, and Exponential Models
Set 2: Geometric Sequences

Station 4

Work with a partner to answer each equation. Use natural logarithms to solve the equations. Use a calculator as needed. Show all your work.

1. If some term a_n of a geometric sequence is 200,000,000, and $a_1 = 2$ and $r = 10$, what is n?

2. If some term a_n of a geometric sequence is 3,145,728, and $a_1 = 3$ and $r = 4$, what is n?

3. If some term a_n of a geometric sequence is 781,250, and $a_1 = 2$ and $r = 5$, what is n?

4. If some term a_n of a geometric sequence is 59,049, and $a_1 = 3$ and $r = 3$, what is n?

5. If some term a_n of a geometric sequence is 1,024, and $a_1 = 4$ and $r = 2$, what is n?

6. If some term a_n of a geometric sequence is 93,312, and $a_1 = 2$ and $r = 6$, what is n?

Geometry

Set 1: Conics

Instruction

Goal: To guide students to facility with the equations and graphs of circles, ellipses, hyperbolas, and parabolas

Common Core Standards

Geometry: Expressing Geometric Properties with Equations

Translate between the geometric description and the equation for a conic section.

G-GPE.1. Derive the equation of a circle of given center and radius using the Pythagorean Theorem; complete the square to find the center and radius of a circle given by an equation.

G-GPE.2. Derive the equation of a parabola given a focus and directrix.

G-GPE.3. (+) Derive the equations of ellipses and hyperbolas given the foci, using the fact that the sum or difference of distances from the foci is constant.

Student Activities Overview and Answer Key

Station 1

Working with groups, students answer questions about the equations and graphs of circles.

Answers

1. $(x-4)^2 + (y-3)^2 = 9$

2.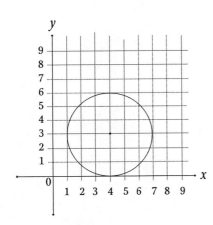

Geometry
Set 1: Conics

Instruction

3. a.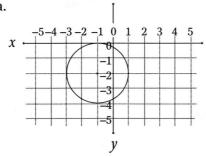

   b. $x = -1$

4.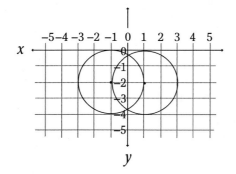

5. $(x-1)^2 + (y+2)^2 = 4$

6. a. 5

   b. $x = 0$

   c. $y = 4$

   d. $y = -1, y = 9$

7.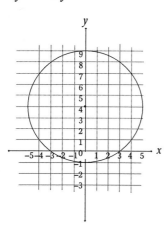

Geometry
Set 1: Conics

Instruction

Station 2

Working with groups, students answer questions about the equations and graphs of ellipses.

Answers

1.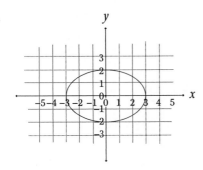

2. a. (0, 0)
    b. 4
    c. (3, 0), (–3, 0)
    d. (0, 2), (0, –2)

3.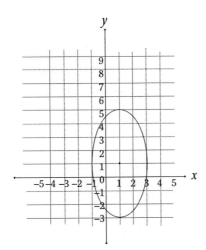

4. a. (1, 1)
    b. 8
    c. (1, 5), (1, –3)
    d. (3, 1), (–1, 1)

Geometry
Set 1: Conics

Instruction

5. (3, 4)

6. $\dfrac{(x-3)^2}{9} + \dfrac{(y-4)^2}{16} = 1$

7.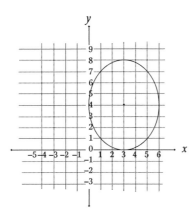

Station 3

Working with groups, students answer questions about the equations and graphs of hyperbolas.

Answers

1.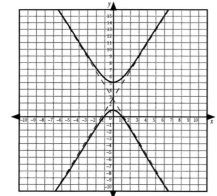

2. $\left(0, 3 \pm \sqrt{5}\right)$

3. (0, 1), (0, 5)

4. a. circles

   b. (4, 2)

5. (4, 2)

Geometry
Set 1: Conics

Instruction

6. $y = x-2, y = 6-x$

7. $\dfrac{(x-4)^2}{2} - \dfrac{(y-2)^2}{2} = 1$

8.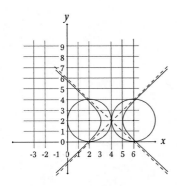

Station 4

Working with groups, students answer questions about the equations and graphs of parabolas.

Answers

1.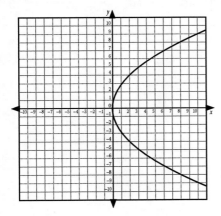

2. $x = -2$

3.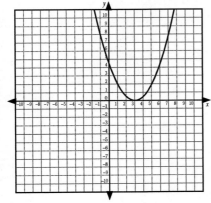

Geometry
Set 1: Conics

Instruction

4. $y = -\dfrac{1}{2}$

5. $(3, 0)$

6. $\left(3, \dfrac{1}{2}\right)$

7.

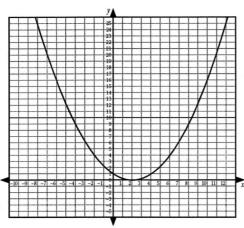

8. By completing the square and writing the equation in standard form, $\dfrac{x^2}{4} + \dfrac{(y+1)^2}{3} = 1$, students should recognize this as an ellipse.

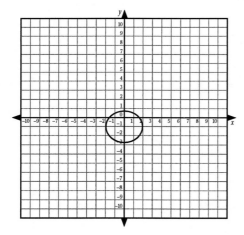

Materials List/Setup

Station 1   colored pens or pencils
Station 2   colored pens or pencils
Station 3   colored pens or pencils
Station 4   colored pens or pencils

Geometry
Set 1: Conics

Instruction

Discussion Guide

To support students in reflecting on the activities and to gather some formative information about student learning, use the following prompts to facilitate a class discussion to "debrief" the station activities.

Prompts/Questions

1. What are conics?
2. What do conics have in common?
3. What is the difference between a circle and an ellipse?
4. What is the relationship between the discriminant and the graph of a conic?

Think, Pair, Share

Have students jot down their own responses to questions, then discuss with a partner (who was not in their station group), and then discuss as a whole class.

Suggested Appropriate Responses

1. Conics are two-dimensional figures that are cross sections of a cone: circles, ellipses, hyperbolas, and parabolas.
2. All conics can be expressed as quadratic equations. All conics are curved.
3. A circle has a single center and radius. An ellipse is a collection of points at a certain distance from two different foci.
4. If a discriminant is less than 0, the graph will be an ellipse, a circle, or a point. If the discriminant equals 0, the graph will be a parabola, two parallel lines, or one uncurved line. If the discriminant is greater than 0, the graph will be a hyperbola or two intersecting lines.

Geometry
Set 1: Conics

Instruction

Possible Misunderstandings/Mistakes
- Incorrectly completing the square
- Incorrectly factoring
- Misunderstanding the relationship between the equation and the graph of a conic
- Mistaking vertices for foci in an ellipse or a hyperbola
- Mistaking the horizontal transverse axis for the vertical, or vice versa, in a hyperbola
- Mistaking the major axis for the minor, or vice versa, in an ellipse
- Not understanding that the radius is perpendicular to any tangent of a circle
- Making arithmetical errors

Geometry
Set 1: Conics

Station 1

Work with your group to answer the questions about each circle.

1. What is the equation of a circle with center (4, 3) that is tangent to the *x*-axis?

2. Graph the circle in problem 1.

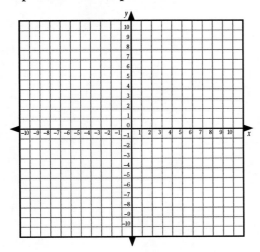

3. Circle *C* has the equation $(x + 1)^2 + (y + 2)^2 = 4$.

   a. Graph the circle.

   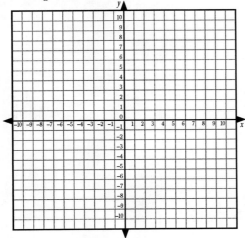

   b. What is the circle's vertical axis of symmetry?

continued

Geometry
Set 1: Conics

4. On the graph in problem 3, use a different color to draw circle D, which has the same radius as C, is tangent to C's vertical axis, and has a center at (1, −2).

5. What is the equation of circle D?

6. A circle has the equation $x^2 + y^2 - 8y + 16 = 25$.
   a. What is its radius?
   b. What is its vertical axis of symmetry?
   c. What is its horizontal axis of symmetry?
   d. What is a horizontal line tangent to it?

7. Graph the circle in problem 6.

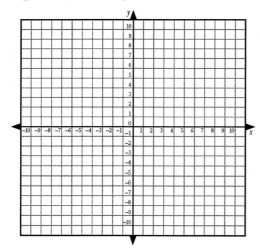

Geometry
Set 1: Conics

Station 2

Work with your group to answer the questions about each ellipse.

1. Graph the ellipse $\dfrac{x^2}{9} + \dfrac{y^2}{4} = 1$.

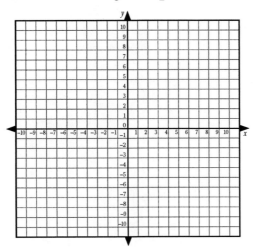

2. Look at the ellipse in problem 1.

   a. What are the coordinates of its center?

   b. What is the length of its minor axis?

   c. What are the coordinates of its vertices?

   d. What are the coordinates of its co-vertices?

3. Graph the ellipse $\dfrac{(x-1)^2}{4} + \dfrac{(y-1)^2}{16} = 1$.

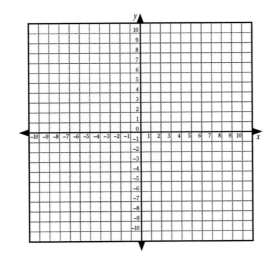

continued

Geometry
Set 1: Conics

4. Look at the ellipse in problem 3.

   a. What are the coordinates of its center?

   b. What is the length of its major axis?

   c. What are the coordinates of its vertices?

   d. What are the coordinates of its co-vertices?

5. If the ellipse C is tangent to the x-axis at a vertex (3, 0) and tangent to the y-axis at a co-vertex (0, 4), what are the coordinates of its center?

6. What is the equation of ellipse C?

7. Graph ellipse C.

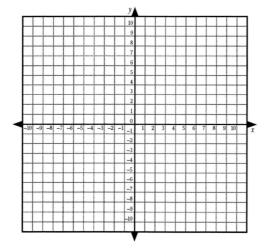

Geometry
Set 1: Conics

Station 3

Work with your group to answer the questions about each hyperbola.

1. Graph the hyperbola $\dfrac{y^2}{4} - \dfrac{3}{2}y + \dfrac{9}{4} - x^2 = 1$.

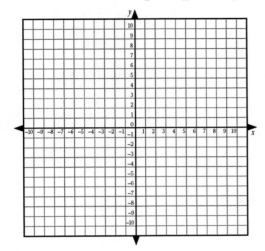

2. Look at the hyperbola in problem 1. What are the coordinates of its foci?

3. Where does the hyperbola intersect the y-axis?

4. Look at the equations $(x - 2)^2 + (y - 2)^2 = 4$ and $(x - 6)^2 + (y - 2)^2 = 4$.

    a. What figures do they represent?

    b. At what point do the figures meet?

5. If the centers of the figures in problem 4 are also the foci of a hyperbola, what are the coordinates of its vertex?

6. The asymptotes of the hyperbola intersect the figures at (2, 0), (6, 0), (4, 2), (2, 4), and (6, 4). What are the equations of the asymptotes?

continued

Geometry
Set 1: Conics

7. What is the equation of the hyperbola?

8. Graph the figures from problem 4 and the hyperbola on the same graph.

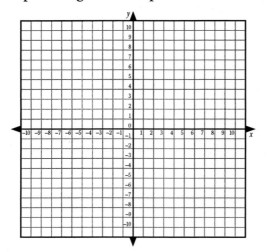

Geometry
Set 1: Conics

Station 4

Work with your group to answer the questions about each parabola.

1. Graph $8x = y^2$.

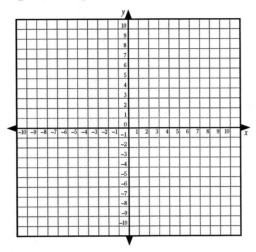

2. In problem 1, what is the parabola's directrix?

3. Graph $2y = (x - 3)^2$.

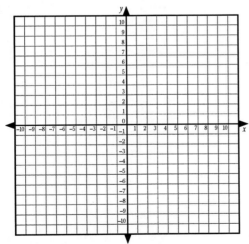

4. In problem 3, what is the parabola's directrix?

continued

Geometry
Set 1: Conics

5. In problem 3, what are the coordinates of the parabola's vertex?

6. In problem 3, where is the parabola's focus?

7. Graph $y = \dfrac{x^2}{4} - x + 1$.

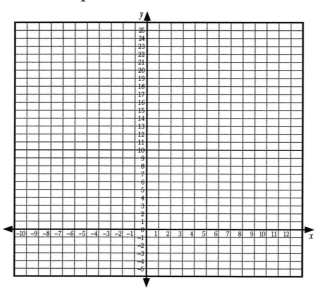

8. Without graphing, predict whether the equation $3x^2 + 4y^2 + 8y = 8$ is a circle, an ellipse, a hyperbola, or a parabola. Graph to check your work.

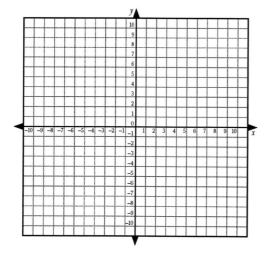

Statistics and Probability

Set 1: Modeling

Instruction

Goal: To guide students to an understanding of modeling, using linear and quadratic regression to analyze data sets

Common Core Standards

Statistics and Probability: Interpreting Categorical and Quantitative Data

Summarize, represent, and interpret data on two categorical and quantitative variables.

S-ID.6. Represent data on two quantitative variables on a scatter plot, and describe how the variables are related.

  a. Fit a function to the data; use functions fitted to data to solve problems in the context of the data. Use given functions or choose a function suggested by the context. Emphasize linear, quadratic, and exponential models.

  c. Fit a linear function for a scatter plot that suggests a linear association.

Interpret linear models.

S-ID.9. Distinguish between correlation and causation.

Student Activities Overview and Answer Key

Station 1

Working with groups, students analyze a data set to find a linear relationship between variables. Students use visual estimates and the median-median line to find the line of best fit.

Answers

1.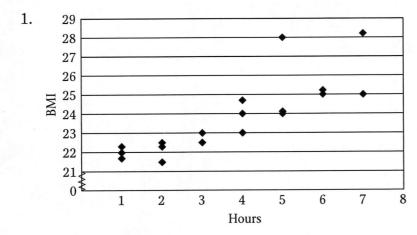

2. yes

Statistics and Probability
Set 1: Modeling

Instruction

3. Answers will vary but should be in the form $y = mx + b$.

4. Answers will vary.

5.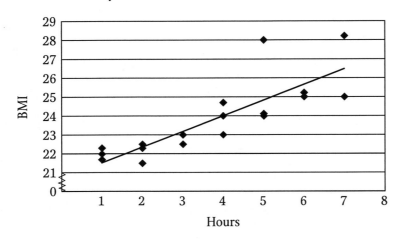

6. No. There are too many other factors involved (such as activity level and diet). There seems to be a correlation, but we can't prove causation.

7. Yes; (5, 28) and (7, 28.2)

Station 2

Working with groups, students analyze a data set to find a linear relationship between variables. Students use a calculator to conduct linear regression.

Answers

1.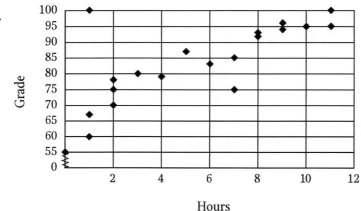

2. yes

3. Answers will vary.

4. (1, 100)

Statistics and Probability
Set 1: Modeling

Instruction

5. $y = 2.67x + 68.64$

6. Yes. The linear relationship is very close.

7. We can't prove causation from this data. As the outlier shows, some people may not study because they already know the material well.

Station 3

Working with groups, students analyze a data set to find a quadratic relationship between variables. Students use a calculator to conduct quadratic regression.

Answers

1.
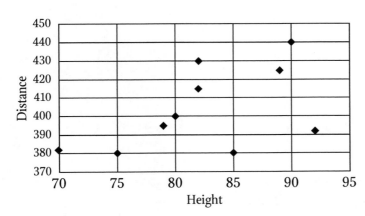

2. yes

3. Answers will vary.

4. Answers will vary, but students should graph a curve resembling a parabola over the points.

5. Answers will vary. Examples include pitch height, pitch speed, wind speed, and direction.

6. There is probably some causative relation, but we can't prove it from this data set.

Statistics and Probability
Set 1: Modeling

Instruction

Station 4

Working with groups, students analyze a data set to find a quadratic relationship between variables. Students use a calculator to conduct quadratic regression.

Answers

1.
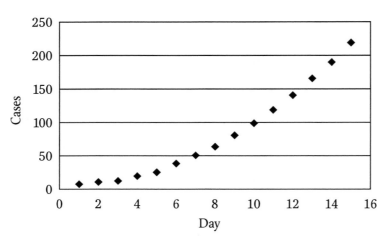

2. quadratic

3. $y = x0.999x^2 - 0.88x + 7.60$

4. Answers will vary, but students should note that deviation is greater in the early days of the epidemic. As the epidemic progresses, the graph is smoother.

5. Answers will vary, but students may suggest that in the early days of an epidemic, cases often go unrecognized and unreported.

6. Answers will vary. Students might suggest that health officials could use the equation to predict the future spread of the disease and make sure that city hospitals order enough vaccine.

Materials List/Setup

Station 1   calculator; colored pens or pencils; graph paper
Station 2   calculator; colored pens or pencils
Station 3   calculator; colored pens or pencils; graph paper
Station 4   calculator; colored pens or pencils; graph paper

Statistics and Probability
Set 1: Modeling

Instruction

Discussion Guide

To support students in reflecting on the activities and to gather some formative information about student learning, use the following prompts to facilitate a class discussion to "debrief" the station activities.

Prompts/Questions
1. What is a scatter plot?
2. What is regression analysis?
3. What is correlation?
4. What is causation?

Think, Pair, Share

Have students jot down their own responses to questions, then discuss with a partner (who was not in their station group), and then discuss as a whole class.

Suggested Appropriate Responses
1. A scatter plot is a graph of many different sets of variables as points on a grid.
2. Regression analysis means attempting to find a linear or quadratic equation that connects the points in a scatter plot and gives a pattern to the data.
3. Correlation is a relationship between variables.
4. Causation is a relationship in which one variable causes the other one to behave in a certain way.

Possible Misunderstandings/Mistakes
- Incorrectly graphing points
- Incorrectly finding medians
- Confusing medians with means
- Incorrectly using the calculator's linear regression and quadratic regression functions
- Confusing correlation with causation
- Incorrectly applying the formula of a line

Statistics and Probability
Set 1: Modeling

Instruction

- Incorrectly applying the formula of a parabola
- Misunderstanding the scenario that provides the data set
- Students with language or culture barriers may find the scenario setups especially confusing. Some students may not be familiar with the concept of home runs in Station 3. Draw the parabolic curve of a home run on the board, and explain that a ball in any sport follows a similar curve through the air. Label the parabola's height at its vertex and its length (the distance the ball travels).

Statistics and Probability
Set 1: Modeling

Station 1

Work with your group to answer each question about the data set. Use the calculator to calculate medians, if needed.

A doctor is trying to find out whether there is a correlation between TV viewing and high body mass. She records average daily viewing habits and takes Body Mass Index (BMI) measurements from 18 people.

TV (hours)	BMI	TV (hours)	BMI
1	22	2	22.3
3	22.5	4	23
5	24.1	6	25.2
7	25	2	21.5
4	24.7	5	28
7	28.2	6	25
5	24	2	22.5
1	22.3	1	21.7
3	23	4	24

1. Graph the doctor's results on a scatter plot. Use graph paper.

2. Does there seem to be a linear relationship between the variables?

3. Estimate the equation for the line of best fit.

4. Draw your line in a different color on the scatter plot.

5. Find the equation for the median-median line. Draw it on the scatter plot in a third color.

6. Can you claim with certainty that increased TV viewing causes higher BMI? Explain.

7. Does the graph have any outliers? If so, what are their coordinates?

NAME:

Statistics and Probability
Set 1: Modeling

Station 2

Work with your group to answer each question about the data set. Use the calculator to calculate medians and create your graphs.

A class wants to find out if there is a correlation between the number of hours studied and grades on the midterm exam. The 20 students log their hours and their grades, as follows.

Studying (hours)	Grade
10	95
1	60
7	75
11	100
1	100
2	70
9	94
7	85
5	87
8	93

Studying (hours)	Grade
2	78
2	75
8	92
3	80
0	55
4	79
9	96
6	83
1	67
11	95

1. Enter the numbers into your calculator to graph the results on a scatter plot. Sketch your plot below.

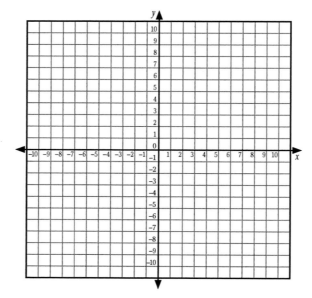

continued

Statistics and Probability
Set 1: Modeling

2. Does there seem to be a linear relationship between the variables?

3. Estimate the equation for the line of best fit.

4. Are there any outliers? If so, what are their coordinates?

5. Use the calculator to find the equation for the line of best fit.

6. Is there a correlation between the variables? Explain.

7. Is there a causative relationship between the variables? Explain.

Statistics and Probability
Set 1: Modeling

Station 3

Work with your group to answer each question about the data set. Use the calculator to calculate medians, if needed.

A baseball team wants to improve its hitting. The team studies the maximum height of home run balls hit at its home stadium and the distance each hit travels. Hits that travel less than 360 feet land within the playing field and can be caught, so the team does not include those hits in its study.

Maximum height (in feet)	Distance (in feet)
80	400
85	380
75	380
82	430
90	440
70	382
82	415
79	395
89	425
92	392

1. Graph the results on a scatter plot. Use graph paper.

2. Does there seem to be a quadratic relationship between the variables?

3. Estimate the quadratic equation for the curve of best fit.

4. Draw your curve in a different color on the scatter plot.

5. What factors besides height might affect the distance of a home-run hit?

6. Is there a causative relationship between home-run height and distance?

Statistics and Probability
Set 1: Modeling

Station 4

Work with your group to answer each question about the data set. Use the calculator to calculate medians and create your graphs.

A health organization tracks the spread of an epidemic through the population of a city, starting with the day on which cases are first observed.

Day	Documented cases
1	8
2	11
3	13
4	20
5	26
6	39
7	51
8	64
9	81
10	99
11	119
12	141
13	166
14	190
15	219

1. Enter the numbers into your calculator to graph the results on a scatter plot. Sketch your plot on graph paper.

2. Does there seem to be a linear or a quadratic relationship between the variables?

3. Use the calculator to find the equation for the line of best fit.

4. What do you notice about points that deviate from the curve?

5. What might be an explanation for these points of deviation?

6. How could the health organization use this information?

Statistics and Probability

Set 2: Sampling

Instruction

Goal: To provide opportunities for students to develop concepts and skills related to examining the five-number summary statistics, mean, standard deviation, random sampling, and mean absolute deviation

Common Core Standards

Statistics and Probability: Interpreting Categorical and Quantitative Data

Summarize, represent, and interpret data on a single count or measurement variable.

S-ID.2. Use statistics appropriate to the shape of the data distribution to compare center (median, mean) and spread (interquartile range, standard deviation) of two or more different data sets.

Student Activities Overview and Answer Key

Station 1

Students are given a graphing calculator. They work as a group to calculate the mean and standard deviation by hand. Then students check their work using a graphing calculator. They are asked to describe the procedure for calculating mean and standard deviation, and to explain what these terms mean.

Answers

1. $\bar{x} = 82$; to calculate the mean, sum the test scores and then divide by the number of scores. In this case, the sum of the test scores is 492 and the number of test scores is 6. Therefore, the mean is 492/6, which is 82.

Statistics and Probability
Set 2: Sampling

Instruction

2. σ = 11.28; to calculate the standard deviation, subtract the mean from each test score. Square each difference and then find the sum of squared differences. Divide that sum by the number of test scores and finally take the square root. Students might find using a table such as the one below helpful in calculating the standard deviation.

x_i	$x_i - \bar{x}$	$x_i(-\bar{x})^2$
77	−5	25
89	7	49
71	−11	121
94	12	144
95	13	169
66	−16	256

Sum the third column, which is 764. Divide by the number of test scores, which is 6. Then take the square root. The standard deviation is 11.28.

$$\sigma = \sqrt{\frac{764}{6}} = \sqrt{127.33} \approx 11.28$$

3. Answers will vary, but the mean should be 82 and the standard deviation should be ≈ 11.28. If students have discrepancies between hand and calculator calculations, there should be some explanation of where the mistake was found.

4. The mean is a measure of center, so the data has a center at 82 points. This is also called the average. The standard deviation is a measure of spread. The standard deviation is ≈ 11.28, so the average difference between any data point and the mean is about 11.28 points.

Station 2

Students use a graphing calculator to find the mean and standard deviation of three classes' test scores. Then they compare the test scores based on the means and standard deviations.

Statistics and Probability
Set 2: Sampling

Instruction

Answers

1. Completed table:

	Mean	Standard deviation
First period	77.8	11.69
Second period	82	7.89
Third period	84.1	9.96

2. The third period class performed the best because they had the highest mean of 84.1 points.

3. The first period class performed the worst because they had the lowest mean of 77.8 points.

4. The first period's scores varied the greatest, as evidenced by the largest standard deviation of 11.69.

5. The second period's scores varied the least because their standard deviation was the lowest at 7.89 points.

6. The third period scored the best of the three classes, with the highest average of 84.1 points. However, the scores varied widely since the standard deviation is almost 10 points. This suggests that even though there were high scores there were also some low scores. The first period class had the worst performance, with an average of 77.8 points, and their scores varied widely with a standard deviation of 11.69. The second period class was in the middle of the performance scale with a mean of 82 points, but their scores had the smallest variation with a standard deviation of 7.89 points. This suggests that perhaps this class had fewer extreme scores.

Station 3

Students work together to explore the concepts of posing questions and performing random sampling. They compare the mean and standard deviation of samples from two populations.

Answers

1. Answers will vary. Sample answer: What is your overall GPA?

2. Answers will vary. Sample answer: Alphabetize the school roster and then separate the boys from the girls. Number the boys list and number the girls list. Use a random number generator or table to choose 100 or more numbers for the boys, and then use the same procedure to choose 100 or more girls. Select the students that correspond to the numbers chosen and pose the question to each student. This is simple random sampling.

Statistics and Probability
Set 2: Sampling

Instruction

3. Answers will vary. If students choose to collect numerical data, the mean and standard deviation might be good measures to use. Be sure that the display matches the type of data collected.

4. Answers will vary. Sample answer: On average, boys seem to study 0.5 fewer hours than girls do, but there are wider variations of study habits in boys than in girls. This is evidenced by the standard deviation for boys (0.67 hours), compared with the standard deviation for girls (0.4 hours).

Station 4

Students are given a number cube to create a data set of 6 points. They calculate the mean and standard deviation using a graphing calculator. Then they are asked to create a second data set of 6 points that has the same mean as the first data set but a different standard deviation. Then they compare the standard deviation and the data sets to explore the variability of data.

Answers

1. Answers will vary depending on the numbers rolled.
2. Answers will vary depending on the numbers rolled.
3. Answers will vary depending on the numbers rolled.
4. Answers will vary depending on the mean and standard deviation calculated in problems 2 and 3.
5. Answers will vary depending on the mean calculated in problem 2. Be sure that the mean is the same as the mean in problem 2.
6. Answers will vary depending on the standard deviation calculated in problem 3. Be sure that the standard deviation is different from the standard deviation in problem 3.
7. Answers will vary. Sample answer: Start with the mean and then choose a number that is 2.5 units above the mean and a number that is 2.5 units below the mean. Then choose two more data points based on the mean. Choose a point that is 5 units above the mean and another one that is 5 units below the mean. Then, for the sixth point, choose the mean again.

Statistics and Probability
Set 2: Sampling

Instruction

8. Answers will vary. Sample answer:

Roll result	Data value
1	−1.5
2	1
3	3.5
4	3.5
5	6
6	8.5

The first data set has values that differ only by 1. The values are closer to the mean than in the second data set. This is supported by the comparison of standard deviations. The standard deviation for the first data set is 1.7, whereas the standard deviation for the second data set is 3.22. The range on the second data set is larger than the range on the first data set.

9. By reporting the standard deviation along with the mean, you get a clearer picture of the distribution. Two data sets could have the same mean but very different-looking distributions. You can only determine how much the data varies about the mean by reporting the standard deviation.

Materials List/Setup

Station 1   graphing calculator
Station 2   graphing calculator
Station 3   none
Station 4   six-sided number cube; graphing calculator

Statistics and Probability
Set 2: Sampling

Instruction

Discussion Guide

To support students in reflecting on the activities and to gather some formative information about student learning, use the following prompts to facilitate a class discussion to "debrief" the station activities.

Prompts/Questions
1. What are the mean and standard deviation?
2. Why is the standard deviation important to report along with the mean?
3. Explain why random sampling is important.
4. Explain how to use the mean and standard deviation to interpret data sets.

Think, Pair, Share

Have students jot down their own responses to questions, then discuss with a partner (who was not in their station group), and then discuss as a whole class.

Suggested Appropriate Responses
1. The mean is a measure of center and is calculated by summing the data values and then dividing by the number of data values summed. The mean is also referred to as the average. The mean is affected by extreme values in the data set. The standard deviation is a measure of spread. The standard deviation is calculated by finding the difference between each data value and the mean, squaring the difference, summing the differences, dividing by the number of data points, and, finally, taking the square root. The standard deviation is an average of how much the data varies from the mean. The larger the standard deviation, the more the data varies.

2. Two data sets can have the same mean but not be identical. The standard deviation tells the audience how widely the data varies from the mean. Without the standard deviation of the data, we don't have a complete picture of the data set.

3. Random sampling increases the chance of selecting data that better represents the population. It gives each subject an equally likely chance of being selected.

4. Since the mean is a measure of center or an average, this number can be used to summarize the data. However, the summary is incomplete without the standard deviation. The standard deviation must be reported along with the mean to indicate how widely the data varies. If the standard deviation is small, then the data values cluster around the mean. If the standard deviation is large, then the data values are widely spread out about the mean.

Statistics and Probability
Set 2: Sampling

Instruction

Possible Misunderstandings/Mistakes
- Forgetting to divide by the number of data points to calculate the mean
- Forgetting to take the square root of the variance for the standard deviation
- Misunderstanding the concept of the standard deviation
- Miscalculating the sums and differences in mean and standard deviation
- Misunderstanding how to obtain a random sample

Statistics and Probability
Set 2: Sampling

Station 1

At this station, you will find a graphing calculator. Work as a group to answer the questions using the following information.

> Mr. Smith teaches five science classes. Each of his classes takes the same final exam. A sample of students from Mr. Smith's first-period class scored the following on the final exam:
>
> 77, 89, 71, 94, 95, 66

1. Calculate the mean of the test scores by hand. Then, describe the procedure used in calculating the mean.

   Mean = _____

2. Calculate the standard deviation by hand. Then, describe how you calculated the standard deviation.

   Standard deviation = _____

continued

Statistics and Probability
Set 2: Sampling

3. Use a graphing calculator to check your work. If your answers do not agree, find and explain the error in your calculations.

   Mean using a calculator = _____

   Standard deviation using a calculator = _____

   Error:

4. What do the mean and standard deviation indicate about the data set?

NAME:

Statistics and Probability
Set 2: Sampling

Station 2

At this station, you will find a graphing calculator. Work as a group to answer the questions using the following information.

Ms. Juarez teaches three math classes. Each class has only 10 students. Each class takes the same midterm exam. Ms. Juarez's classes scored the following on the midterm exam:

First period	Second period	Third period
77	89	98
81	77	94
67	94	86
91	74	81
56	74	87
95	71	83
88	79	76
84	91	89
72	90	87
67	81	60

1. As a group, find the mean and standard deviation of each data set using a graphing calculator. Fill in the table below.

	Mean	Standard deviation
First period		
Second period		
Third period		

continued

Statistics and Probability
Set 2: Sampling

2. Which class performed the best on the test? Justify your answer.

3. Which class had the worst test performance? Justify your answer.

4. Which classes' scores varied the greatest? Justify your answer.

5. Which classes' scores varied the least? Justify your answer.

6. Write a summary comparing the scores for the classes. Include in your description each mean and standard deviation.

Statistics and Probability
Set 2: Sampling

Station 3

Use the situation below to answer the questions.

Several articles have been published claiming that boys are better in school than girls. Recently published articles suggest that girls perform better in school than boys do. You want to determine what the trend is in your school.

1. What question would you pose to your schoolmates to determine whether boys or girls perform better in school?

2. Assume that it is not possible to question each student. Describe what sampling method you would use.

3. What statistics or graphical displays would you use to present your data?

Alejandro and Bea found that, in their school, girls perform better than boys do. They wanted to investigate the reason for this. They asked their schoolmates in a random sample how many hours on average they study every night. The results are below.

	Mean (hours)	Standard deviation
Boys	1.3	0.67
Girls	1.8	0.4

4. Use the mean and standard deviation to explain the results.

NAME: _____

Statistics and Probability
Set 2: Sampling

Station 4

At this station, you will find a number cube. You will create a set of data with the number cube. Find the mean and standard deviation, then create your own data set with the same mean but a different standard deviation.

Work as a group to create the first data set by rolling a six-sided number cube 6 times.

1. Record the results in the table below.

Roll	Result
1	
2	
3	
4	
5	
6	

2. What is the mean of your data set? _____

3. What is the standard deviation of your data set? _____

4. Create a second data set with the same mean but a different standard deviation than your first data set. Use 6 numbers. You can use any real numbers, including negative numbers.

Data point	Value
1	
2	
3	
4	
5	
6	

5. What is the mean of your second data set? _____

continued

Statistics and Probability
Set 2: Sampling

6. What is the standard deviation of your second data set? _____

7. Describe how you created your second data set.

8. Compare the data sets by first writing each set in numerical order in the table below. What do you notice?

Roll result	Data value

9. Explain why the standard deviation must be reported along with the mean.

Statistics and Probability

Set 3: z-scores

Instruction

Goal: To guide students to an understanding of z-scores in relationship to standard deviation, probability, and intervals

Common Core Standards

Statistics and Probability: Interpreting Categorical and Quantitative Data

Summarize, represent, and interpret data on a single count or measurement variable.

S-ID.4. Use the mean and standard deviation of a data set to fit it to a normal distribution and to estimate population percentages. Recognize that there are data sets for which such a procedure is not appropriate. Use calculators, spreadsheets, and tables to estimate areas under the normal curve.

Student Activities Overview and Answer Key

Station 1

Working with groups, students calculate the z-score from a given mean and standard deviation.

Answers

1. $z = \dfrac{13 - 14.54545}{1.50756} = -1.0251$

2. $z = \dfrac{6 - 5.875}{1.24642} = 0.1003$

3. $z = \dfrac{70 - 56.81818}{21.82576} = 0.60396$

4. $z = \dfrac{20 - 21.5}{1.28602} = -1.16639$

5. $z = \dfrac{32 - 30.9}{1.44914} = 0.75907$

Statistics and Probability
Set 3: z-scores

Instruction

Station 2

Working with groups, students calculate the z-score from a given set of data and standard deviation. Students should begin to develop a sense of the relationship between z-scores and real-world data.

Answers

1. a. 14.625

   b. 3.2838

   c. $z = \dfrac{16 - 14.625}{3.2838} = 0.4187$

2. a. 75.95238

   b. 13.06704

   c. $z = \dfrac{90 - 75.95238}{13.06704} = 1.07504$

   d. $z = \dfrac{65 - 75.95238}{13.06704} = -0.83817$

3. a. 94.8

   b. 5.13809

   c. $z = \dfrac{100 - 94.8}{5.13809} = 1.01205$

   d. $z = \dfrac{89 - 94.8}{5.13809} = -1.12882$

4. a. 49.8

   b. 4.36654

   c. $z = \dfrac{60 - 49.8}{4.36654} = 2.33594$

Station 3

Working with groups, students use z-scores to determine probability.

Answers

1. a. 19.84615

   b. 2.33973

   c. 1.35

   d. 0.0885

   e. 8.85%

Statistics and Probability

Set 3: z-scores

Instruction

2. a. 37.5 d. 0.0212
   b. 2.71038 e. 2.12%
   c. −2.03

3. a. 13.6 d. 0.2358
   b. 4.99333 e. 23.58%
   c. −0.7210

Station 4

Working with groups, students use z-scores and probability to determine the approximate number of items in a particular interval.

Answers

1. 25.17%
2. 30.51%
3. 8.19%
4. a. 0.33%
   b. 12.79%
   c. The shift between midnight and 8 A.M. should get first priority.
   d. Answers will vary.

Materials List/Setup

Station 1   calculator
Station 2   calculator
Station 3   calculator; z-scores table
Station 4   calculator; z-scores table

Statistics and Probability
Set 3: z-scores

Instruction

Discussion Guide

To support students in reflecting on the activities and to gather some formative information about student learning, use the following prompts to facilitate a class discussion to "debrief" the station activities.

Prompts/Questions
1. What is probability?
2. What is standard normal distribution?
3. How do we graph standard normal distribution?
4. How do we designate an interval within standard normal distribution?

Think, Pair, Share

Have students jot down their own responses to questions, then discuss with a partner (who was not in the same station group), and then discuss as a whole class.

Suggested Appropriate Responses
1. Probability is the mathematical likelihood of an occurrence.
2. Standard normal distribution refers to a function that shows the distribution of occurrences across an interval, with a mean of 0 and standard deviation of 1.
3. Standard normal deviation is graphed with a bell curve whose area equals 1.
4. We designate an interval within standard normal distribution by shading part of the area beneath the bell curve.

Possible Misunderstandings/Mistakes
- Incorrectly setting up the z-score calculation or incorrectly calculating the z-score
- Misinterpreting the data
- Misreading the z-scores table
- Incorrectly using a calculator's statistics function
- Subtracting z-scores rather than probabilities to find the probability
- Rounding too soon in the series of calculations

Statistics and Probability
Set 3: z-scores

Station 1

Work with your group to find each z-score. You may use a calculator to find the z-score. Show how you used the formula.

1. $M = 14.54545$

   $s = 1.50756$

   What is the z-score for a sample value of 13?

2. $M = 5.875$

   $s = 1.24642$

   What is the z-score for a sample value of 6?

3. $M = 56.81818$

   $s = 21.82576$

   What is the z-score for a sample value of 70?

4. $M = 21.5$

   $s = 1.28602$

   What is the z-score for a sample value of 20?

5. $M = 30.9$

   $s = 1.44914$

   What is the z-score for a sample value of 32?

Statistics and Probability
Set 3: z-scores

Station 2

Work with your group to answer each question. Use a calculator to find the standard deviation. You may use a calculator to find the z-score. Show how you used the formula.

1. At a track meet, students land long jumps of the following distances, in feet:

   10, 12, 13, 15, 20, 19, 18, 15, 15, 16, 17, 12, 11, 9, 14, 18

   a. What is the mean?

   b. What is the standard deviation?

   c. Find the z-score that represents the likelihood of a student jumping at least 16 feet.

2. On a recent test, students received the following grades:

   90, 75, 76, 55, 40, 92, 88, 80, 80, 81, 78, 65, 67, 92, 89, 72, 78, 83, 84, 62, 68

   a. What is the mean?

   b. What is the standard deviation?

   c. Find the z-score that represents the likelihood of a student scoring at least a 90.

   d. If a failing grade is anything below 65, find the z-score that represents the likelihood of failing.

continued

Statistics and Probability
Set 3: z-scores

3. A food company is trying to create a new shipping crate for their most popular apple. Apples that weigh 89 grams or less will not be desirable to consumers. The new shipping crate will not accommodate apples of 100 grams or more. Apples outside the weight range are used for apple sauce, juice, and frozen pies. The apples in a sample have the following weights, in grams:

    100, 90, 85, 92, 93, 95, 98, 99, 94, 102

    a. What is the mean?

    b. What is the standard deviation?

    c. Find the z-score that represents the likelihood of an apple weighing 100 grams or more.

    d. Find the z-score that represents the likelihood of an apple weighing 89 grams or less.

4. Job applicants receive the following scores on a skills aptitude test:

    50, 45, 49, 47, 48, 52, 53, 60, 48, 46

    a. What is the mean?

    b. What is the standard deviation?

    c. Find the z-score that represents the likelihood of an individual's scoring higher than 60.

Statistics and Probability
Set 3: z-scores

Station 3

Work with your group to solve the problems about z-scores and probability. Use a calculator and z-scores table.

1. At a track meet, shot-put competitors record the following distances, in feet:

    20, 19, 18, 20, 21, 23, 22, 15, 16, 20, 21, 22, 21

    a. What is the mean?
    b. What is the standard deviation?
    c. Find the z-score that represents the likelihood of a shot-put distance over 23 feet.
    d. What is the corresponding number on the z-scores table?
    e. What is the probability that someone will record a distance over 23 feet?

2. To determine whether he can safely plant a crop, a farmer records the following overnight low temperatures, in degrees Fahrenheit:

    40, 43, 39, 35, 34, 37, 39, 36, 41, 38, 38, 35, 36, 34

    a. What is the mean?
    b. What is the standard deviation?
    c. Find the z-score that represents the likelihood of the temperature falling below freezing (32°F).
    d. What is the corresponding number on the z-scores table?
    e. What is the probability that the weather will turn freezing overnight?

3. The same farmer has a record of April rainfalls for the past 10 years, in inches:

    20, 15, 3, 17, 11, 10, 19, 16, 13, 12

    If April rainfall is less than 10 inches, the farmer must plan extra watering.

    a. What is the mean?
    b. What is the standard deviation?
    c. Find the z-score that represents the likelihood of rainfall less than 10 inches.
    d. What is the corresponding number on the z-scores table?
    e. What is the probability that the farmer will have to plan extra watering?

Statistics and Probability
Set 3: z-scores

Station 4

Work with your group to solve the problems about z-scores, probability, and intervals. Use a calculator and z-scores table.

1. Amy learns that on this morning's history exam, the mean of students' scores was 75.66667, with a standard deviation of 11.19216. There are 30 students in the class. What is the probability that Amy's score is between 80 and 90?

2. This afternoon, another class of 30 students will take the same test. What is the probability that a student will receive a score between 65 and 75?

3. A medical association conducts a study on cholesterol levels in men in a certain population. If the mean cholesterol is 200, with a standard deviation of 39.8, what is the probability that a man in this population will have cholesterol between 170 and 180?

4. A city needs to allocate its traffic officers to the times of day when traffic violations are most likely to occur. It assigns each reported violation a number corresponding to the time of day the violation occurred, between 1 and 24, with 1 representing 1 P.M., 2 representing 2 P.M., and so on, all the way up to 24, or noon. The city nicknames these numbers "time stamps." After several months a pattern becomes clear. The mean time stamp of traffic violations is 13.5, with a standard deviation of 3.

    a. What is the probability that a violation will occur between the hours of 5 P.M. and 6 P.M.?

    b. What is the probability that a violation will occur between the hours of 2 A.M. and 3 A.M.?

    c. Given the standard employment shifts of 8 A.M. to 4 P.M., 4 P.M. to midnight, and midnight to 8 A.M., when should the city plan to have the most traffic officers on call?

    d. Is the city's method of assessing its data on traffic violations a valid approach? Explain.